About the author

Robert Cohen is Emeritus Professor of Medicine in the University of London, UK. He was educated at Plymouth and Clifton Colleges and spent most of his career at The London Hospital (now The Royal London Hospital) and its Medical College. His main interests were internal medicine and metabolic disorders, especially those of the liver and kidneys. He is the author of many papers, and the writer or editor of several multi-authored volumes on these topics. He is married to Professor Barbara Boucher and they have two children and five grandchildren. He and his wife retired in 1999, and now live in a small village a few miles north of Chichester, West Sussex, UK.

LADDERS

The History and Science of Elevation

Robert D. Cohen

Emeritus Professor of Medicine
St. Bartholomew's and The London
School of Medicine and Dentistry
Queen Mary University of London, UK

Matador
9 Priory Business Park,
Wistow Road, Kibworth Beauchamp,
Leicestershire. LE8 0RX
Tel: (+44) 116 279 2299
Fax: (+44) 116 279 2277
Email: books@troubador.co.uk
Web: www.troubador.co.uk/matador

ISBN 978 1783063 253

British Library Cataloguing in Publication Data.
A catalogue record for this book is available from the British Library.

Printed and bound in the UK by TJ International, Padstow, Cornwall
Typeset in 11pt Adobe Garamond Pro by Troubador Publishing Ltd, Leicester, UK

Matador is an imprint of Troubador Publishing Ltd

Contents

Introduction   vii

Physical Aspects   1

Metaphorical Aspects   77

Notes   117

Index   121

**Plates
Indicates an entry in the Notes
**All photographs are taken from WikiMedia Commons*
[August 2013]

Introduction

There are two sorts of ladders – the physical ones, and those which represent human striving to get to the top, by fair means or foul. The latter I have called 'metaphorical'. There is inevitably a certain amount of mathematical material in the 'physical' section, but the non-mathematically orientated reader can conveniently ignore this sort of stuff. I have provided a glossary of terms used, which may be helpful to some. In some cases I have given references to source material. Since ladders are means of getting up and down, I have also included accounts of other devices serving the same purposes, such as ropes, staircases, escalators and lifts (elevators, for US readers!), and bridges, which are means of maintaining one's height, when one might otherwise be falling into the abyss!

Chapter 1

Physical aspects of ladders

Everyone is familiar with ladders; there can scarcely be a household in the country which doesn't boast one or more. The main problem is storage, since ladders are large and not visually attractive. Many keep them in their roof spaces or garages, and some, notably building contractors, in their vans or in racks on the roofs of their cars. In order to achieve length, many ladders are built in sliding sections, so that they maybe overlapped when not in use. Kitchens often contain small ladders, better referred to as 'steps', so that the upper storage shelves in larders or cupboards may be reached.

Ladders have rungs to climb on and usually have parallel sides, though the top section may be tapered. In this respect they represent either convergent or infinite series. A series is an orderly sequence of numbers, with each number having a logical relationship to its predecessor and successor. A simple example is the series

1, 4, 7, 10, 13, 17, 21, 24... etc.

in which each number is separated from its neighbour by three. If such a series continues indefinitely, and has no finite sum, then it is divergent. If an infinite series has a finite sum, then it is known as 'convergent'.

To give an example, in a finite arithmetic series the n-th term is

$$x_n = x_1 + (n\text{-}1)d$$

where d is the common difference between the terms. The sum of a finite series of n terms is:-

$$S_n = (n/2) \cdot (x_1 + x_n)$$

We can illustrate this by calculating the sum of all even whole numbers from 2 to 100 (i.e. 50 numbers).

$$S_{50} = (100/2).(2 +100)$$

i.e. `$S_{50} = 2550$

Another more modest form of ladder is the kitchen steps, used for retrieving cans from the top shelf of one's larder. Typically, the majority of cans are baked beans, usually of the Heinz variety, though there are somewhat less satisfactory versions from many other sources.

H.J. Heinz was born in Pittsburgh, Pennsylvania, on 11th October, 1844 (I am deeply honoured to share a birthday which such a distinguished person!). He was the son of German immigrants to the USA and set up a small food produce business, which grew into the Heinz Corporation, founded in 1876. He was educated at Duff's Business College and died in 1919. There are a great number of Heinz canned foods other than baked beans – famously, '57 varieties'. These varieties are listed in Wikipedia under 'Heinz 57'and, in addition to the more well-known items, such as baked beans, cream of mushroom soup, tomato ketchup and mock turtle soup, include some less well known delicacies such as Heinz Gumbo Creole, Heinz Spanish Queen Olives and Heinz Chow Chow pickle. These days there are substantial competitors, e.g. The Big Soups, which are full of chunks of meat and vegetables, and I have a sneaking preference for these. I had thought that this preference might be due to the type of preservative used, but both Heinz and The Big Soups state that they contain 'no artificial preservatives'.

One of the most dramatic types of ladder is that used by firemen, either to hose water on to conflagrations on the upper floors of buildings, or to rescue trapped personnel. These ladders are an integral part of fire engines and are highly extendable. The problem I have always had is to work out why two or three firemen in the business end of such a ladder don't manage to overbalance the whole set up. There must be a massive counter-weight in the carriage of the fire engine; part of this, of course, must be the water reservoir, but as this gets used up, there must be something else stabilising the situation. When firemen are not otherwise occupied, they may be summoned to rescue cats that have got stuck up trees, up which they have dashed to in order to escape pursuing dogs, or merely out of sheer bloody mindedness! The rescuing fireman is usually rewarded for his pains by receiving long scratches down the side of his face!

An important protection for firemen is that provided by asbestos, which is resistant to fire, heat, electrical and chemical damage. When used for resistance to fire or heat, the asbestos fibres are often mixed with cement, or woven into fabric or mats. Asbestos is obtained from serpentinite rocks, which are widely found throughout the world. A common form is chrysotile. Its use dates back at least 4,500 years, according to evidence from East Finland, where it was employed to strengthen earthenware cooking pots. The word *'asbestos'* is derived from a Greek word meaning 'unquenchable' or 'inextinguishable'. Unfortunately, in the early 1900's it was noted that there were a large number of early deaths and lung problems in asbestos-mining towns, and in the building and marine construction industries (which used asbestos for pipe-lagging). These were mainly due to fibrosis of the lungs (asbestosis), and two types of cancer – ordinary carcinoma of the lung, and an otherwise rare condition known as mesothelioma, which is a cancer of the pleura (the lining of the lung and interior wall of the chest). These conditions are caused by the inhalation of asbestos fibres. In Australia, the use of asbestos in the construction industry was banned in 2003. In the UK, the use of blue and brown asbestos materials in factories and offices was banned in 1985. The owners of factories have a legal duty to manage previously installed asbestos lagging, removing it if necessary. Employers whose staff may come into contact with asbestos must provide annual training to its staff on how to handle it. There are now a number of substitutes for asbestos, such as fibre-glass insulation and glass wool. A major incident involving asbestos was the Al Qaeda attack on the World Trade Center buildings on 2001, when terrorist-piloted aircraft crashed themselves into the twin towers of the Center releasing much asbestos fibre used for insulation. Asbestos can be rendered harmless by heat treatment at 1000-1250°C, which converts it into silicate glass.

Assault ladders are used for boarding ships or capturing buildings in combat operations. In mediaeval times they were placed against the wall of enemy fortresses. The soldiers who were unfortunately designated to make the first assault were usually met by a deluge of boiling oil or molten tar, and the defenders would make strenuous efforts to upset the ladders, thereby often killing, or seriously harming, those climbing up. Various means of overwhelming the defenders were used. These included the fire arrow. which had a wad of oil soaked material attached to it, the idea being to set the wooden houses within the fort alight and thereby distract the defenders. The trebuchet was a mediaeval version of the sling or catapult used in Christian and Muslim lands around the Mediterranean Sea in the 12th century. It could fling projectiles weighing up to 350 pounds at high speeds into enemy fortifications but became obsolete in the 15th century, well after the invention of gunpowder. This weapon was constructed from a long beam pivoted on an axle. On one side of the axle there is the short arm of the beam, attached to a heavy weight. The long arm of the beam can be pulled down so that some offensive projectile can be attached to its end. This manoeuvre raises the heavy weight on the short arm. The end of the short arm is then released, and the stored energy in the counterweight is transferred efficiently to the projectile, which flies towards the target.

It seems to me that the trebuchet may have been the forerunner of the modern mortar bomb, which also lobs unpleasant objects into the heart of the enemy's redoubt. Mortars have existed for many centuries, it being claimed by many historians that they were first used in 1453 in the siege of Constantinople by the Turks. Somewhat more recently they were used by Government forces to suppress the 1719 Jacobite rising at the battle of Glen Shiel. We have a watercolour of this site and have also visited it. A tall pillar nearby commemorates the event. Mortars were used in the American Civil War at the siege of Vicksburg by forces under General Ulysses S. Grant. All these mortars were very heavy, non-portable devices, and it was not until World War I that a version transportable by a single person was produced. These were particularly useful in the muddy ground of the Western Front, since their trajectory allowed them to fall into the enemy's trenches. They were used by the Provisional Irish Republican Army (IRA) to attack British forces in the last decades of the twentieth century. In the Newry mortar attack of 1985, nine members of the Royal Ulster Constabulary were killed. In 1991, the IRA used mortars to attack 10, Downing Street. Fortunately only one bomb detonated – it landed in the back garden of the Prime Minister's residence, but the only damage done was to the windows. However, the Prime Minister (John Major) had to move to temporary accommodation in Admiralty House whilst repairs were effected.

There are many other sorts of ladders. In our previous house in Beckenham we had a loft ladder. There was a trap-door in the roof and, when this was opened, a ladder could be slid down to allow access to the roof. Ladders with hooks on the upper end are used to grip window sills, and are also employed by fire-fighters. External roof ladders have hooks at the top to grip the ridge of steep roofs. Orchard ladders usually have three legs, and as, their names implies, they are used for picking fruit.

Staircases are of two main types – ordinary linear ones, which may ascend buildings by repeated flights at right angles to each other, and spiral ones. My old place of work, The London Hospital, had five floors, including the ground floor. The wards in which I had patients were, needless to say, on the top floor, which meant about twelve flights, each containing about ten or fifteen steps. Thus only the fittest young doctors and nurses could get to the top without getting pretty exhausted. There were, indeed, elevators, but these were frequently *hors de combat* for one mechanical or electrical reason or another. Sometimes they were full of domestic staff going on or off duty, so that one had to keep fit to answer emergency calls to the wards on the top storeys of the building, because it was impossible to get into the lifts; one had to race up the stairs to the relevant ward, where equipment such as cardiac defibrillators were already in place. In our earlier days, these machines had to be brought up the service lifts, but later each ward was equipped with its own apparatus. One had to be careful not to be in contact with the patient when the defibrillator was activated if one didn't want to get a tremendous electric shock!

The spiral staircase is often a feature of old castle turrets and lighthouses, since they don't take up much space. They have some serious disadvantages, mainly because the steps are roughly triangular in shape and taper towards the central pillar. This makes it difficult when someone else is climbing up at the same time as you are going down, or *vice versa.* You always have the dilemma as to whether you should stick to the outside, or hug the middle, and as soon as you meet the person going in the opposite direction the problem becomes acute! Most spiral staircases have bannisters on the outside, or a rope winding down the central pillar to hang on to. Good examples of accessible spiral staircases are at Castle Gorm in North Mull in Scotland, and in Smeaton's Tower on Plymouth Hoe. The latter is the old version of the Eddystone Lighthouse, brought inland when the present version was built. The new one may be seen from the Hoe on clear days, about fourteen miles out to sea. Sometimes you can just make out the stump of the old Smeaton's Tower in the distance, standing next to the present lighthouse.

One of the most uncomfortable spiral staircases that I know is in the Leaning Tower of Pisa, in which the curious posture of the tower has been created by land subsidence on one side. As you ascend this tower, which has no bannisters or ropes, you feel as if you were climbing up at one moment and down at the next. The tower has been in some danger of toppling over, and the local authorities have always been at great pains to stabilise it, most recently because of its enormous financial value as a tourist attraction; however, all such efforts seem to have deliberately maintained some degree of tilt, perhaps so that its world renown can be maintained. The tower is in fact the campanile (bell tower) of the nearby cathedral, and is about 56 metres high; with about 295 steps; it weighs about 16,200 tons. Before the shoring-up operation, the angle of the deviation from the vertical was 5.5 degrees; this was reduced to 3.99 degrees, implying that the top of the tower is displaced horizontally by 3.9 metres. The tower was built in the last 25 years of the 12th century and subsidence began as soon as the second floor was complete. A large number of construction engineers were employed over the 195 years of the building operations. One likely major player was Diotosalvi, this attribution being related to the fact that he did much other work in the same region and some of the other local buildings have a similar style.

The tower of Pisa has been used for numerous purposes over the centuries since its construction. Galileo Galilei is said to have dropped cannon balls of different weights from its top to show that their speed of descent was independent of their weight. During World War II, the Allies discovered that the Germans were using it as an observation post, but, apparently, a U.S. Army sergeant was so impressed by its beauty that he refrained from ordering an artillery strike on it.

In 1964, Italy requested international aid and expertise to prevent the tower from toppling yet again, and a multi-disciplinary group of historians, engineers and mathematicians assembled in the Azores (I have no idea why they didn't meet in Pisa!) to plan a course of action. They seemed to be taking rather a long time in their deliberations, but impetus was suddenly provided in 1989 by the collapse of the Civic Tower of Pavia. The bells were temporarily removed from the Leaning Tower to reduce the weight and cables were slung around the third level and anchored several hundred metres away. 38 cubic metres of soil were then removed from below the raised end of the tower and it was then straightened by 45 centimetres at the top. The tower was declared 'stable' for another 300 years and was re-opened to the public in 2001. There are several other notable leaning towers world-wide, notably the Capital gate building in Abu Dhabi, the Leaning Tower of Suurhusen and the

14th century bell tower of Bad Frankenhausen, both in Germany. Curiously, in New Zealand the Leaning Tower of Wanaka was deliberately built at an angle of 53 degrees to the ground. I have no idea why – perhaps some quirk of modern architecture.

I must now come to the important subject of library steps. Since the top shelves in many libraries are often out of reach, all such libraries must have some form of ladder to access them. The library of The Royal Society of Medicine at 1, Wimpole Street, London has excellent steps, with a bannister to hold on to in order to prevent one overbalancing with a pile of books in one's arms. When Barbara and I constructed a library in our present home near Chichester, we scoured the country for some library steps, but completely failed to find any, even having searched for two years. I was therefore forced to construct something myself, which I did by cannibalising some kitchen steps and screwing a broomstick in a vertical position to the side of them that we could grab hold of, and this has been very successful.

Gravity

Basically the purpose of all ladders is to overcome gravity, which is the attractive power between two bodies. The Law of Gravity was first stated in quantitative terms by Isaac Newton in his work *Philosophiae Naturalis Principia Mathematica*, first published in 1687. If there are two bodies of masses m_1 and m_2, separated by a distance of r, then the force F between them is given by:-

$$F = gm_1m_2/r^2$$

(Here g is a universal gravitational constant equal to approximately 6.674×10^{-11} Nm^2kg^{-2}.) Thus the attraction is inversely proportional to the square of the distance between the two bodies. The point from which the distance is measured on each body is its centre of gravity. In the case of the bodies being spherical, the centre of gravity corresponds to the centre of the sphere.

The actual origin of gravitational attraction is much more obscure. Einstein produced a plausible explanation by saying that space and time were one and the same (time is known as 'the fourth dimension'). He suggested that gravity was simply a curvature in space-time created by a mass in the same way as a piece of cloth would be curved if it was stretched out and a heavy object placed on it. This curvature in space created by an object with greater mass than another nearby object on the cloth would cause the smaller object to fall towards the greater one. At least this seems plausible, but I have been unable to find an actual mathematical derivation of $F = gm_1m_2/r^2$. It maybe that this law is purely empirical, i.e. derived purely from experimental observation.

It is well-known that Newton dropped a few apples during his deliberations (he was also hit by them as they fell off the tree). He may also have dropped a clanger, since in 1666 Robert Hooke published his thoughts on the subject, which got pretty close to Newton's, though he did not give a mathematical formula for the phenomenon at the time. He later (in January 1679 or 1680) sent to Newton his "supposition that the Attraction always is in duplicate proportion to the Distance from the Center Reciprocall". There was then a fairly gentlemanly academic squabble between the two as to priority, reminiscent of some more modern controversies!

Ladders in stockings

All women who wear stockings or 'pop socks' are beset by the problems of 'ladders' developing, which make their appearance all too often and make their wearers look somewhat sleazy or impoverished. All ladders are initiated by a small hole, caused, for example, by a nick with a sharp finger nail. The trouble is that such ladders are self-propagating – once they have started there is little chance of the ladder failing to progress all the way down the leg to the foot. There are, however, now available several brands of ladder-resistant stockings. In these, there is far less likelihood of the ladder propagating from the initial nick. It seems that none of the types of stockings, e.g. silk or nylon is immune to the development of ladders. Silk stockings were much prized during World War II and were virtually unattainable except from American soldiers. I recall GI's prowling the streets of Plymouth during the War, offering the local girls pairs of silk stockings in return for certain favours which it is unnecessary to detail!

Walking under a ladder and bad luck

The origin of the belief that walking under a ladder brings bad luck is uncertain, but there are a number of theories. One derives from the early days of Christianity, and is related to the concept of the Trinity, i.e. that God is made up of three parts – the Holy Father, the Holy Son and the Holy Ghost (or Spirit). This made the number 3 an object of veneration. A ladder leaning against a wall creates a triangle, with the ladder as the hypotenuse, with the wall and ground at right angles to each other forming the other two sides. To walk through this triangle, i.e. walking under the ladder, was seen as breaking up the Trinity and was therefore considered to be blasphemous and the perpetrator was then considered to be in league with the Devil! Another explanation was that a ladder leaning against a wall had some similarity to a gallows, and gallows, for obvious reasons, were considered to be emblems of bad luck. There were many prescriptions for warding off the evil effects of walking under a ladder; these included spitting three times through the rungs of the ladder, or spitting on your shoe. There was an important condition to allow the shoe-spitting to be effective – you must make sure you didn't look at your shoe until the spit has dried! If these remedies don't work, you must cross your fingers till you find a dog, or back out the way you came and make a wish! Another somewhat more bizarre predictive procedure is to feed a snake whilst stroking a cat! This last manoeuvre is clearly for those on jungle safaris, whilst, for unknown reasons, carrying a ladder!

Jacob's ladder (Plate 1(a))
This is described in Genesis 28, verses 10-19. Jacob was sleeping on a journey between Beersheba and Haran, when he dreamt of a ladder set up on earth and climbing up to Heaven, where were assembled God and his angels. God identified himself and declared that the descendants of Jacob would cover the face of the Earth. When Jacob awoke, he said "Surely the Lord is in this place and I did not know it". Jacob became afraid and said "This is none other than the house of God" and named it '*Bethel*', which is the ancient Hebrew for 'the house of God' ('*Beth*' is the Hebrew for 'house' and '*el*' means God). There are numerous interpretations by later authors of the meaning of the dream. One of the most coherent is that the ladder signifies a bridge between Heaven and Earth and solders a connection between God and the Jewish people; the story is thus one of the origins of the belief that the Jews were God's chosen people, thereby condemning the Jews to centuries of persecution by those (often known as the Gentiles) who objected to the Jews arrogating God to themselves.

Other luminaries had other interpretations of the dream, e.g. St. John Chrysostom who stated:-

"And so mounting as it were by steps, let us get to heaven by a Jacob's ladder. For the ladder seems to me to signify in a riddle by that vision the gradual ascent by means of virtue, by which it is possible for us to ascend from earth to heaven, not using material steps, but improvement and correction of manners". In another interpretation, by Adam Clarke, an early 19th century Methodist theologian and scholar, Christ himself is seen as the ladder bridging the gap between Heaven and Earth. A 7th century monk at the monastery of Mount Sinai, St. John Climacus, wrote a tract ('The Ladder of Divine Ascent'), describing how to raise one's body and soul to God through the acquisition of ascetic virtues. His ladder has thirty steps, the first seven concerning general virtues needed for the ascetic life, the next nineteen containing guidance on overcoming vices, and the final four steps describe the higher virtues to which the ascetic aspires. The final step of all was love. (e.g. The Beatles – *'All you need* is *love'!).* In the square outside Edgware Road Underground Station in London is a rather good statue of a man carrying a ladder; I have no idea who the sculptor was (Plate 1(b)).

It should be noted that the concept of Jacob's ladder is not confined to the Christian and Jewish faiths. Jacob is also an Islamic prophet and patriarch and Muslims regard the ladder as symbolic of the need to follow 'the straight path'. The twentieth century scholar Martin Lings interprets the Islamic view of the Ladder as follows;-

"The ladder of the created Universe is the ladder which appeared in a dream to Jacob, who saw it stretching from Heaven to Earth, with Angels going up and down upon it; and it is also the 'straight path', for indeed the way of religion is none other than the way of creation itself retraced from its end back to its Beginning", though I find all this pretty incoherent! There is another version of Jacob's ladder in a painting by William Blake, in his inimitable style, with a spiral staircase representing the ladder, being ascended by numerous willowy females swathed in semi-transparent clothing (Plate 1(a)). In the Catacomb Via Latina in Rome, there is a wall painting by Marsyas, in typical Roman style (Plate 1(c)). At the entrance to Bath Abbey, on the north side of the entrance is a ladder stretching from ground level to the top, with numerous figures climbing up it, no doubt on their journey to Heaven (Plate 1(d)). Albrecht Dürer painted a *'Deposition from the Cross',* which included a ladder; it had a figure of Christ showing the Stigmata, and surrounded by numerous horrified but sympathetic women (Plate 2(a)). In Chatsworth House, Derbyshire, a glazed porcelain tower, by Felicity Aylieff, decorated with ladders, adorns the entrance hall (Plate 2(b)); Aylieff was born in 1954 and studied ceramics at the Bath Academy of Art. In 1889, Vincent van Gogh painted a picture entitled *'Een boerin die vlas kneust (naar Millet)',* the meaning of which escapes me, despite the use of a Dutch dictionary and phrase book! (Plate 2(c)); it shows a woman from the back, probably threshing corn manually, with a ladder, of unknown significance, at the right. Plate 2(d) is of a *Nativity*, by Piero della Francesca, in which all the people depicted are very solid, with their feet very much on the ground, in contrast to the painting shown in Plate 3(a) – *'The Promenade'*, done by Marc Chagall in 1917.

'Honi soit qui mal y pense'
This is the motto of the Order of the Garter and was originated by King Edward III who was dancing with his first cousin and daughter-in-law Joan of Kent. When one of her garters slipped down to her ankle, causing the attendant courtiers to snigger at her humiliation. The King removed the offending garter and placed it around his own leg, saying *'honi soit qui mal y pense'*, which may be translated from the French as 'A scoundrel, who thinks badly by it'. A more modern schoolboy translation is 'Honey, your silk stocking's hanging down!' Other translations are:-

"Shame be to him who thinks evil of it" (from the Old French)
and:-
"Evil be to him who evil thinks".

Knights and Ladies of the Garter are entitled to encircle their heraldic arms with a band containing this inscription (i.e. the Old French version). The Royal Coat of Arms of the United Kingdom contains this later inscription (not the schoolboy translation!), together with the motto *'Dieu et mon droit'*. Many public buildings, e.g. law courts display these mottos and they are stamped in gold on the front of British passports, and on the stationery of companies operating under Royal Warrant. The British Army also uses it extensively – for instance the Horse Guards have the motto stamped on their ceremonial helmets. The motto is used also by many Commonwealth countries – e.g. by the Australian, New Zealand and Canadian Armies, the Canadian Royal Regiment and the Canadian Grenadier Guards. As for *'Dieu et mon droit'*, this is the motto of the British Monarch, and refers to the divine right of the monarch to govern. It was first used by Richard the Lionheart as a battle cry, and was adopted as the royal motto of England by King Henry V in the 15th century. Richard the Lionheart had the common medieval belief that victory in battle went not to the side with the better army, but to that which God viewed with the most favour. It has been translated in several ways – e.g. 'God and my right', 'God and my lawful right' and 'God and my right shall me defend'. Needless to say, it has been lampooned – thus the Beatles pop group paraphrased it as 'Duit on Mon Dei' ('Do it on Mondays')! As in the case of '*Honi soit qui mal y pens*', it appears on the front page of the British passport.

Michelangelo's frescoes on the ceiling of the Sistine Chapel
Anyone seeing these masterpieces must wonder how Michelangelo (1475-1564) accessed the ceiling after he was commissioned by Pope Julius II in 1508 to repaint the ceiling, on which the previous decoration had been merely golden stars on a blue sky. He was at first very reluctant to take on the task, being suspicious that the whole thing was a plot set up by his enemies in the hope that he would disgrace himself by failing (or, perhaps, falling!). Michelangelo initially turned down the commission since he saw himself as a sculptor rather than a painter! The task has previously been attempted by the noted architect, Donato Bramante (1444-1514), who was responsible for the design of St. Peter's Basilica in Rome. He had perforated the vault in order to lower strings to support the scaffold. This was frowned on by Michelangelo, since it would leave the holes in the vault once the work was finished. The Pope ordered Michelangelo to build a scaffold of his own, and he did this by creating a flat wooden platform supported by wall brackets, which he reached by using a ladder.

Amongst the most famous images in the Sistine Chapel is the outstretched Hand of God giving life to Adam. There are also nine paintings of stories from the Book of Genesis. These include depictions of the Creation, Adam and Eve in the Garden of Eden and the Great Flood. The total area of the frescoes is over 5,000 square feet. Michelangelo was, of course, also responsible for 'The Last Judgement', on the wall behind the altar. This was painted between 1535 and 1541, and is a depiction of the second coming of Christ and the Apocalypse. Christ is judging individual souls, and, according to his verdict, they then either ascend to Heaven, or descend to Hell. The fresco gave rise to considerable controversy because all the figures were naked, and the artist was accused of immorality and obscenity. The so-called Fig-Leaf Campaign was organised by Cardinal Carafa and Monsignor Semini, Mantua's ambassador to the Pope, whose Master of Ceremonies, Biagio da Cesena declared that it was disgraceful that in so sacred a place there should have been depicted all those nude figures, exposing themselves so shamefully, and "that it was no work for a Papal Chapel but rather for the public baths and taverns". The Pope wisely retorted that his jurisdiction did not extend to hell, so the work would have to remain!

This sort of prudery was not, of course confined to the Renaissance. Amongst the most modern advocates of 'decency' was the redoubtable Mary Whitehouse (1910-2001), who attacked the British media – e.g. the British Broadcasting Corporation – for encouraging a more permissive society. Surprisingly, she was a teacher of sex education! She became involved in evangelical movements such as the Student Christian Movement and Moral Rearmament. She was the leading light in the 'Clean-Up TV' group, and founded the National Viewers' and Listeners' Association, using it as a platform to criticise the BBC for "excessive portrayals of sex, violence and bad language" and attempts to "remove the myth of God from the minds of men". She instigated a number of court cases, accusing various people and bodies of obscenity, a number of which she actually won. She started a private prosecution against '*Gay News*'. Needless to say, she was extensively lampooned for her prudery. Summing all this up, one might think that she was rather a pain in the neck! Nevertheless, the tendency to be sexually explicit in the media is once more being criticised, especially for the way in which it might affect the young – so maybe Mrs. Whitehouse is being vindicated!

Aviation

Since the dawn of Man's history it has been his ambition to emulate the birds and get into the air – by this I mean travel laterally through the air for considerable distances, as opposed to just jumping up and down, or leaping over a gap. So it seems appropriate to start with the evolution of flight in birds. It is obvious that the smaller the weight of the creature, the easier it will be for it to fly. Birds have evolved bones which are filled with air sacs. Teeth have been replaced by beaks, which are made of *keratin, a lighter material than that of the constituents of teeth. Because of the absence of teeth, masticatory activity has to take place in the bird's gizzard. The whole problem of the mechanism of flight engaged the attentions of Aristotle and other Greek philosophers in about 350BC. About 150 years ago, a fossil of the ancestral bird *Archaeopterix* was discovered, and much theoretical speculation arose from this event. In 1999 a 'pouncing Proavis' model was proposed by Garner and colleagues. This model suggested that birds evolved from predators specializing in ambush from elevated sites. Amongst other virtues, the authors thought that this model predicted the sequence of character acquisition in avian evolution, and explained the evolution of feathers, beginning with simple forms that produced a benefit by increasing drag. In 1879, Williston, and later Ostrom[2], described the use of wings as an insect-foraging mechanism, which later evolved into a wing stroke. This model proposes that flight emerges in running bipeds through a series of short jumps. Ostrom compared the amount of energy expended by various hunting models with the amount of food gathered and showed that the potential food haul can be doubled in this way. Thus, to gather the same amount of food, *Archaeopteryx* would expend less energy in jumping than by running alone, so the cost/benefit ratio would be greater. However, this theory was subject to considerable criticism. In order to achieve lift-off, *Archaeopteryx* would have had to run three times faster than modern birds, because of its weight. Since the mass of the bird and the distance needed to run in order to achieve lift-off are proportional, the energy needed to take off increases, and, once airborne, the drag due to wind resistance would slow the flight speed and the bird would thus be forced to land again. Another hypothesis was that primitive birds ran up tree trunks to escape predators, and presumably glided down on their wings when the danger was over – but this met with all sorts of objections. The so-called 'arboreal model' considered *Archaeopteryx* as a reptile that soared from tree to tree and, as it did so, it would use its primitive wings as a balancing mechanism. Furthermore, the conservation of energy during the gliding phase would give this model advantages in energy efficiency over other models. The flying squirrel is possibly a modern equivalent. Obviously, the whole question of the evolution of bird flight it still very much up in the air!

The actual shape of the wing is a big factor in determining the types of flight possible for a particular species of bird. An important parameter is the 'aspect ratio', which is the ratio of the square of the wingspan to the wing area. 'Elliptical wings' are short and rounded and allow for tight manoeuvring, e.g. in hawks. 'High speed' wings are short and pointed. This sort of wing is seen in peregrine falcons, which have dive speeds of up to 200 mph. The fastest level flight is seen in certain swifts and may reach 105 mph. 'High aspect ratio' wings are used for slower flight and for dynamic soaring (kestrels, terns, nightjars, and birds that specialise in soaring and gliding flight, e.g. stormy petrels above the waves at sea). The ability to fly at low speeds is important in birds that plunge for fish. Eagles and vultures have slots between the feathers at the ends of the wings, both to reduce drag and to capture the energy in the air flowing from the lower to the upper surfaces of the wing. The ability to hover is another specialization, seen for example in hummingbirds and some kingfishers; this is obviously a highly energy-consuming form of flight: some hummingbirds have been observed by cinematographic methods to beat their wings up to 52 times per second. (I have had the experience while fishing for trout of a kingfisher perching on the end of my rod. Since the bird almost immediately dived and emerged with a fish in its beak, whereas I caught nothing all day, I deduced that the bird must have been cheating!)

Snakes and ladders

This is just about the first board game that children are introduced to, commonly at the age of four or five years, and is possibly their first introduction to the concepts of good and evil in the world, the ladders representing 'good' and the snakes 'evil'. The game is of Indian origin, and is known in that subcontinent as '*Moksha Patam*'; one version, developed by the Jain sect and known as '*Gyanbazi*' dates back to the 16th century. *Moksha Patam* was a derivation of traditional Hindu philosophy, contrasting *karma* (destiny) and '*kama*' (desire). The ladders represent virtues (e.g. generosity, faith and humility) and the snakes indicate vices such as lust, anger, violence and theft. The reason why the game board has fewer ladders than snakes is symbolic of the fact that the path of virtue is much more difficult to tread than the path of sin. There are obvious similar overtones in western religions – thus many of our clerics are anxious to point out that we are all primarily sinners, and after death have to spend varying periods in Purgatory before admission by St. Peter through the Pearly Gates to the empyrean heights of Heaven and the presence of God. In the Roman Catholic branch of Christianity, the process is aided by the practice of confession of sins and the administration of absolution by the priest hidden in the confessional and by their presence at deathbeds.

The Jewish religion confines its apologia for sins largely to the Day of Atonement ('*Yom Kippur*'), which is a day of deprivation of food and water, a regime imposed to aid concentration on one's shortcomings. During the *Yom Kippur* service, forgiveness is requested for just over fifty types of sin, of which some are in the mind and others which have been actually committed. To give a few examples, in the first category are:-

Hardening of the heart, envy, levity, being stiff-necked.

There are far more in the second category, e.g.:-

Lack of chastity, wanton glancing, swearing, wronging a neighbour, despising parents and teachers, violence, blasphemy, chattering foolishly, usury, cheating in business, gorging at meals, drinking too much alcohol, telling tales.

On *Yom Kippur*, the whole recitation of sins ends up with requesting pardon for the sins for which we deserve the four kinds of death which can be ordered by a Court of Law – stoning, burning, beheading and strangling! Civilisation has clearly progressed somewhat since this spine-chilling prescription was first produced, but clearly the early leaders of Israel had to have some weapon with which to keep order!

Rope ladders (see below under 'Ropes')

Ladders in Art

Because of the symbolic aspects of ladders, they are frequently depicted in paintings. The illustrations in this book give a number of examples. The earliest is that painted on the walls of the catacombs in the Via Latina, Rome. There is the typical lack of perspective, the techniques for which had to wait until the time of the Renaissance. The history of the development of perspective in art is described by Manetti (1480) and Giorgio Vasari (1550), the latter being famous for his biographical work 'The Lives of the Most Eminent Painters, Sculptors and Architects'. Although Vasari (1511-1574) was no mean painter and architect himself, he was scarcely in the first rank, and is most famous for his book. Nevertheless, he built the loggia of the Uffizi Palace, and built the passage (the Vasari Corridor), which connects the Uffizi with the Pitti Palace on the other bank of the River Arno. This enclosed corridor crosses the Ponte Vecchio and winds around the outside of many buildings. Vasari also renovated the churches of Santa Maria Novella and Santa Croce. He built the octagonal dome on the Basilica of Our Lady of Humility in Pistoia and painted an *'Adoration of the Magi'*, commissioned by Pope Pius V. Vasari was the first to use the word 'Renaissance' *(rinascita)* in print. His biographies are full of amusing stories, an example being that the young Giotto painted a fly on a painting by Cimabue, which the latter repeatedly tried to brush away! Cimabue (1251-1302) was the last well-known Italian artist who painted in the Byzantine style, i.e. with no perspective and with the background and halos in gold. The name 'Cimabue' may be translated as 'bull-headed', because he instantly destroyed anything in his atelier which he didn't like, and for this reason Dante placed him in Purgatory in the Divine Comedy, (though the length of the sentence that Dante thought he should serve is not clear). Cimabue's best known work today is the Crucifix painted for the Church of Santa Croce in Florence, a work which was largely destroyed in the floods which occurred in 1966, traces of which may still be found today in that city.

Cimabue's best known pupil was Giotto (Giotto di Bordone, 1266-1337). The best known story about him was that the Pope of the day got wind of his skill and sent a messenger to him asking him to demonstrate his skill by sending one of his drawings to him. Giotto drew a circle which was perfect, despite not using the aid of a compass, and this so impressed the pope that he was summoned to work in Rome.

A century or so later, the formidable Albrecht Dürer (1471-1528) was in full flow in Nuremberg. A gallery in Dresden contains his famous Deposition; in this work, there is a conspicuous ladder behind the Cross, with its tip resting on the sign 'INRI' (Jesus of Nazareth, King of Israel). Dürer's works usually have a characteristic device consisting of a capital 'A', beneath the cross-stroke of which nestles the 'D', but this is not obviously present in this painting. However, the *Crown of Thorns is lying on the ground in the bottom left hand corner, and this seems to be designed so as to represent an 'A' and 'D'. Dürer, apart from being a painter of both oils and watercolours, was also an engraver, printmaker and mathematician. Amongst his treatises are considerations of perspective and of ideal proportions. He was taught the principles of perspective in Bologna, by either Luca Pacioli or Bramante. He also learnt how to construct shadows, following a technique of Leonardo da Vinci. This last piece of information is taken from the entry on Dürer in Wikipedia, but I am not quite sure what it means. It may be referring to the length of the shadow, which, of course, varies with both latitude and the time of day; alternatively, it may allude to the fact that shadows are best represented in art by a deep mauve or purple, rather than by black or brown. Dürer was interested in the variations in bodily proportions seen in men and women, and wrote '*Vier Bücher von Menschlicher Proportion*' ('Four Books on Human Proportion). He based these ideas on Vitruvian Man, the concept developed by Leonardo da Vinci, which showed how the whole body, with outstretched limbs, could be inscribed within a circle. What the significance of this concept was is obscure, at least to me!

An interesting contrast in styles is provided by Piero della Francesca (c.1415-1492) (Plate 2(b)) and Marc Chagall (1887-1985) (Plate 3(a)). In the former, all the human figures appear to have real weight, with their feet very firmly planted on the ground. In contrast, Chagall's figures are frequently floating about in the air, often accompanied, for some reason, by donkeys! This curiosity was apparently inspired by the Bible and by stories he was told as a child (though what these stories were is unclear).

Aviation

As indicated above, Man has always wanted to emulate the ability of birds to fly. The earliest attempts were made by one or other varieties of the hot air balloon. The brothers Montgolfier were the inventors of this method. Joseph-Michel (1740-1810) and Jacques-Étienne (1745-1799) were born into a family of paper manufacturers in the Ardèche region of France. Joseph was a maverick and dreamer. Étienne was more down-to-earth and interested in the latest Dutch innovations. Joseph was watching a fire one night and observed the sparks and embers arising upwards. He thought that this could help an assault on the fortress of Gibraltar (as much a bone of international contention then as it is today!). He believed that the rising of the smoke and sparks was due to a special gas with a property he called 'levity'. So he and his brother made a box-like chamber and covered the top and sides with lightweight taffeta; they then lit some tinder under it, and the whole contraption rose up and hit the ceiling! They scaled up their model and the lifting force was so great that they lost control of the contraption, which eventually landed about two kilometres away, where it was immediately destroyed by some alarmed passers-by!

In 1783 the brothers staged a public demonstration to establish their claim to its invention in front of a number of influential French dignitaries. The flight lasted about 8 minutes and reached an estimated altitude of 1600-2000 metres. Since there was some concern about the possible effects of altitude, the King of France (Louis XVI, accompanied by Queen Marie-Antoinette), who witnessed the demonstration, proposed that they should launch two criminals, but it seems that the first passengers were actually a sheep, a duck and a rooster. The scientific methods associated with the design of trials were obviously fairly well understood, since the sheep was supposed to have a physiology rather like that of Man, and the birds, which were used to high altitudes, acted as controls. The next step was manned flight and it was Étienne who made the first ascent, the balloon being tethered to the ground. In 1766, the British scientist, Henry Cavendish used hydrogen to fill balloons, rather than rely on the hot air current arising from a fire, and hydrogen was used for many subsequent early ballooning adventures, e.g. the crossing of the English Channel by the Franco-American aviators Jean-Pierre Blanchard and John Jeffries. Hot air ballooning remains a present day sport, there being a centre for this activity just a few miles from our home. We are always afraid that one of these contraptions will wreak havoc by landing in our garden!

Airships

These were the next developments in flight technology. What these had that simple hot air balloons did not, was a means of propulsion, including propellers, and also steering mechanisms in the form of rudders, so that the passengers were not purely dependent on the vagaries of the air currents as to where and when they landed; thus the first airships were often known as 'dirigibles'. The gas filling the balloon was at first hydrogen, but concerns as to inflammability and the danger of explosion led to this being replaced by helium. Airships, of course, derive their flight capability from the large bag of lighter-than-air gas which is their most striking feature. In some airships ('blimps') this bag is non-rigid, so the shape and lifting power relies entirely on the internal pressure of the gas in the balloon. Others have a rigid framework, which may contain many gas-filled cells. This has the advantage that if one of the gas cells ruptures, the buoyancy and lift is maintained by the others.

Airships with rigid frameworks were known as 'Zeppelins', after their inventor, Count Zeppelin. Ferdinand Adolf Heinrich August Graf von Zeppelin (1838-1917) was the son of a Württemberg Minister and Hofmarschall. He attended a polytechnic at Stuttgart and then became a military cadet at Ludwigsburg. He studied science, engineering and chemistry, and in 1863 travelled as an observer for the Union troops with the Army of the Potomac during the American Civil War. Later, in the Franco-Prussian War of 1870/1871, he distinguished himself in a reconnaissance mission behind enemy lines, and this made him famous throughout Germany. From 1882-1885 Zeppelin was a commander of Uhlans, but blotted his copy book during military manoeuvres and was forced to retire from the Army. He then invested in a company which designed and built airships. (There was actually a row about whether he used someone else's designs for his airships). The first of these, the LZ1, was tested near Lake Constance and remained airborne for about 20 minutes, but it was damaged on landing. The next version, the LZ2, was flown, after many mishaps, in 1906, but ran into trouble with a stiff breeze, when the engines failed because of cooling problems, and it was brought down in the Allgau mountains. The third version, LZ3, was rather more successful, and made two flights, achieving forward speeds of up to 36 mph. LZ4 first flew in 1908 and excited so much public interest that a collection campaign raised over 6 million German marks. The airship was now considered sufficiently safe for German royalty to make such flights, and Zeppelin was awarded the Order of the Black Eagle! By 1914 the German Aviation Association transported 37,250 people in over 1600 flights without incident. As an afterthought,

UK readers might like to know that the British rock group Led Zeppelin was once threatened with legal action over the use of the Zeppelin name by Zeppelin's granddaughter, Countess Eva von Zeppelin! The use of the airship declined as that of the aeroplane increased and, in addition, their demise was hastened by a number of accidents, including that of the hydrogen-filled *Hindenberg.* This took place during an attempt to dock at its mooring mast in Lakehurst, New Jersey, USA. Out of the 97 people on board, 35 were killed after the airship burst into flames, possibly ignited by static electricity.

Recently, (August, 1913), the airship has taken on a new lease of life. Gennady Verba, president of the company Augur Inc., has unveiled a model of the *Atlant,* a hybrid craft combining elements of airship, helicopter and aeroplane design. (See The Times, 31st August, 2013), which is expected to make its first actual flight in about three years' time. The design sketches show an aircraft resembling a gigantic cycling helmet, with a hard outer shell and a passenger and cargo area that covers the base of the vehicle. Small and large versions of the craft are planned, carrying 16 and 60 tonnes of cargo respectively. The larger will be 130 metres in length, with a range of 3,750 miles and a maximum speed of 124 mph. Verba's partner, Igor Pasternak, has already flown a similar hard-shell craft, the Aeroscraft Dream Dragon, and has predicted that there will soon be thousands of like craft in the skies. Several other firms are joining the bandwagon – Lockheed Martin said that its new craft would become a 'pick-up truck in the sky'. The British company, Hybrid Air Vehicles, has won a share of a US defence contract worth $500 million for a military defence vehicle. Though the public lost confidence in airships as a result of the Hindenberg disaster of 1937, modern airships are safer, since they are filled with helium (which is non-inflammable), rather than with hydrogen. Verba sees the following main uses for the airships of then future:-

1) Supplying remote communities such as oil and gas workers.
2) Disaster relief work where the infrastructure (such as airports) does not exist.
3) Transporting troops for the military.
4) Luxury travel. "We can create a flying yacht that can accommodate ten to twelve people with a unique level of comfort".

But what happens during a hurricane, typhoon, and hail and snowstorms? My guess is that you get blown about like a leaf in the wind.

All of these previous flight 'machines' were lighter than air, and it was this property that provided their lifting power. The aeroplane was the first heavier-than-air machine to fly, pioneered by the brothers Wilbur (1871-1948) and Orville (1867-1912) Wright. The Wright brothers were two of seven children born to Milton Wright (of English and Dutch descent) and Susan Koerner (of German/Swiss origins). Orville dropped out of school prematurely to start a printing business with Wilbur's help. They then opened a cycle sales and repair shop and used the profits to further their interest in flight. Amongst other things, they noted that birds in flight banked in order to change direction, so they experimented with twisting the wings for this purpose. ('wing warping'), using unmanned models with cords to adjust the angle of the wings. They also curved ('cambered') the top surface of the biplane model used, this giving better lift than a flat surface. It was evident that wings with a higher 'aspect ratio' (*vide supra*) provided better lift and less drag than the broader wings. The Wrights used the Lift Equation:-

$$L = kSV^2.C_L$$

where L is the lift in pounds, k the coefficient of air pressure (Smeaton coefficient), S the total area of lifting surface in square feet, V the velocity (headwind plus groundspeed) in miles per hour and C_L the coefficient of lift (which varies with wing shape). They used a wind tunnel to test their calculations, and introduced a tail rudder that had a flexible portion with which to steer, and they also designed a new type of propeller. They commissioned an engine made of aluminium blocks to keep the weight down. This engine was fed with fuel by gravity feed from a fuel tank mounted on a wing strut, and developed 12 horsepower. The first successful flights were made on 17th December, 1903, firstly with Orville as the pilot, being one of only 120 feet, lasting 12 seconds, but the second and third attempts covered 175 and 200 feet respectively, at an altitude of about 10 feet! The fourth flight, however, made all of 852 feet. Eventually, Wilbur made a flight of 24.5 miles, only landing when the fuel ran out. The brothers had several crashes, fortunately escaping with only minor injuries. It is remarkable how the efforts of these two brothers have entirely changed the whole face of global civilization. Nevertheless, their efforts were met at the time by considerable scepticism. There was obviously a widely held opinion that the photographs taken by the brothers were fakes, and in 1906 the Paris edition of the Herald Tribune headlined an article 'FLYERS OR FAKES?' But, after a demonstration in Paris, the president of the Aéro-Club de France finally admitted that he had done them an injustice. There was the usual flurry of law suits over the priority of the patents for various improvements to the aeroplane. (I sometimes think that the world was invented for the benefit of lawyers!)

There then followed a number of landmark flights. In 1909 Louis Blériot (1872-1936) became the first person to fly across the English Channel in a heavier-than-air machine, thereby winning the prize of £1000 that had been offered for this feat by the *Daily Mail* newspaper. His list of other inventions started with that of a practical headlamp for motor cars; he set up a business selling them, used the profits to build an aircraft, and designed and flew the first practicable monoplane. Blériot had no compass to guide his cross-Channel flight, merely following the shipping visible down below. But visibility deteriorated, and only the appearance of the grey line of the English coast stopped him from getting entirely lost. He eventually landed in a field near Dover Castle; he was taken to the harbour and recuperated in the Lord Warden Hotel (for non-UK readers, there is a ceremonial office, once occupied by Sir Winston Churchill, entitled 'Lord Warden of the Cinque Ports'. The five ports in question are Hastings, New Romney, Hythe, Dover and Sandwich). The Confederation of the Cinque Ports was initially set up for military and trade purposes, but is now no longer in existence. A Royal Charter of 1155 established the ports with the duty to maintain 57 ships for the Crown in case of need. In return, the ports received the following privileges:-

'Exemption from tax and tallage, right of soc and sac, tol and team, blodwit and fledwit, pillory and tumbril, infangentheof and outfangentheof, mundbryce, waifs and strays, flotsam and jetsam and ligan'.

'Tallage' is a land use or land tenure tax, brought to England by the Normans as a feudal duty. It was condemned in Magna Carta (1215) and abolished in England in 1340. 'Soc and sac' refers to full cognisance of all criminal and civil cases within their jurisdiction. 'Tol' was the right granted to a landowner to impose a payment on the passage of goods or cattle through his lands, and to be exempt from the tolls of others. 'Team' was the permission to hold courts to judge people for wrongful possession of goods or cattle. 'Blodwit and fledwit' was the authority to punish shedders of blood and to seize those who flee from justice. The pillory and tumbril refer to punishments for minor offences. 'Infrangentheof and outfrangeneof' are the powers to detain and to execute felons. 'Mundbryce' was the authority to try people for breaches of the

King's Peace. 'Waifs and Strays' was the right to seize lost or unclaimed goods after one year and one day. 'Flotsam and jetsam or ligan' refers to the power to claim floating wreckage on the shore or goods thrown overboard.

The next great landmark in aviation was the crossing of the Atlantic. This was achieved by Charles A. Lindbergh (1902-1974) in an aircraft named '*The Spirit of St. Louis*'. He took off from Roosevelt Field near New York and flew northeast along the coast over Nova Scotia and then over St. John's, Newfoundland, using only a magnetic compass, an airspeed indicator and a good deal of luck. He crossed the Irish coast, and passed over England, landing at Le Bourget Field near Paris, where he was greeted by an enthusiastic crowd of over 100,000 people; his flight had taken 33.5 hours. He rather blotted his copy book when he was invited by the governments of France and Germany to tour their aircraft industries. In 1938 he received a German Medal of Honour from no less a person than Reichsmarshal Herman Goering. For this, he was heavily criticised by those in the USA who took a dim view of Nazism and also for his refusal to hand the medal back in spite of the uproar. He at first actively campaigned against American involvement in World War II, but changed his mind after the Japanese attack on the American base at Pearl Harbour. He had accused British, Jewish and pro-Roosevelt groups of leading America into war, but, having changed his mind, he flew about 50 combat missions. After the war he was employed by Pan American World Airways ('Panam') as a consultant and helped design the Boeing 747 jet. He joined the conservation movement and was especially concerned with the protection of humpback and blue whales. He also opposed the development of supersonic air transport because he feared it might have adverse effects on the Earth's atmosphere. It is interesting to note that the eventual demise of the Anglo-French Concorde was due to other reasons, including not only noise and supersonic bangs, but also a fearsome crash in Paris. I made one trans-Atlantic trip on Concorde, completely unaware of this background, and enjoyed myself greatly. Since I was travelling with a senior official from the UK Department of health and Social Security, I was due to travel on a routine Boeing 747 flight in the first class accommodation (such were the privileges of the UK Civil Service at that time). Suddenly the girl behind the desk in the First Class departure lounge announced rather shamefacedly that the 747 had taken off without calling the First Class passengers. There was absolute uproar, especially from the business tycoons who needed to arrive in New York in sufficient time to clinch important business deals! Matters only settled down when we were offered a flight on Concorde. On this trip they served one of my favourite clarets (Chateau Léoville Lascases, of a particularly outstanding vintage) and steak done in exactly the way you prescribed. Unfortunately, the only East Coast airport which could accommodate Concorde was at Philadelphia, whereas my luggage had already been loaded on to the 747 and was on its way to New York. However, so embarrassed were British Airways about the whole affair, that I was very swiftly re-united with my baggage.

The only form of aviation which I haven't yet experienced is space travel. I don't mean trips to the Moon (which doesn't seem to me to be really worth visiting), but journeys in which it might be quicker to traverse the globe via space, rather in the upper atmosphere. This brings me to the subject of round-the-world trips, at least one of which I have made during a lecture tour. This sort of journey is not very pleasant. Apart from sheer boredom, one suffers from repeated episodes of 'jet-lag'. If you are forced into this sort of activity, the best thing to do is to program repeated stops of at least three days into your schedule – but you are somewhat at the mercy of those inviting you. I remember some curious episodes in Japan; here, when you meet an acquaintance in the street you both bow to each other; careful observation shows that one of the bowing couple usually bows lower than the other. This is because the lower one's social status relative to that of the person one is bowing to, the lower you have to bow! I was also once at a formal dinner in Japan, at which each guest had a *geisha girl standing behind him, attending to all their needs. One of their more dangerous habits was to refill your glass of 'saki' after each sip, so you have very little idea of how much you had actually drunk (until you tried to stand up!). The important guy sitting next to me at one stage threw his head back, went to sleep and started to snore loudly. I was initially horrified at this striking display of bad manners, but later learnt that you were expected to do this to express your appreciation of the meal. Another odd habit of the Japanese is the use of the word 'honourable' at every conceivable occasion. Thus, after giving a lecture, I was accosted by two female students, who, whilst bowing almost to the floor, declared how honoured they were to be present at my honourable lecture!

The barrage balloon

Anyone old enough to have lived through World War II will be familiar with the barrage balloon. Large numbers of these were floated near favourite sites for enemy air raids, and were intended to force the marauding aircraft to fly higher, thus reducing the accuracy of bombing. Enemy aircraft therefore made strenuous attempts to shoot them down. In the UK they were typically manned by members of the Home Guard (otherwise known as 'Dad's Army', since they were usually people who were too old to be sent to the front). Barrage balloons were intended to defend against dive-bombers flying at heights up to 5,000 feet and thus beyond the range of concentrated anti-aircraft fire. Anti-aircraft guns could not traverse quickly enough to attack low flying aircraft. Of course, barrage balloons were ineffective against higher flying aircraft, but they did have some effect against the V-1 flying bomb, and it was officially claimed that 231 V-1s were destroyed by balloons. Many bombers had devices on the leading edge of their wings which triggered a small explosive charge that cut the balloon cable. There were, however, downsides to balloon defence. During severe storms in 1942 some balloons broke loose, and the trailing mooring cables short-circuited power lines, and caused disruption to mining and manufacturing. Balloons have also been used for tests on nuclear weapons, mainly to determine the optimal height at which to explode them. I am happy to say that nowadays they are mostly used for advertising purposes. Since we live near Goodwood race-course, the local balloons are largely used to advertise a local bookmaker called 'Betfair'!

The aerofoil

This is the crucial feature of wing design which provides lift to an aircraft. Looking at a section through the wing parallel to the fuselage, the upper surface has a characteristic curve, more marked at the front of the section. The under surface, by contrast, is uncurved. An aerofoil moving through air or fluid produces a force, the component of which perpendicular to the direction of motion is known as 'lift', whereas the component parallel to the direction of motion is termed 'drag'. The disturbance of airflow caused by the movement of the aerofoil creates curved 'streamlines' which results in lower pressure on the upper side of the wing compared with the lower side, and, obviously, if the lifting forces exceed the drag, the plane ascends. Whilst aerofoils for subsonic planes have a rounded leading edge, in contrast, those designed for supersonic flight have a very sharp leading edge. Some aircraft have wings whose configuration can be altered in flight by the pilot, e.g. from a position at right angles to the fuselage to a more swept-back state. Such aircraft are known as 'swing-wing'. This was a feature of the American General Dynamics F-111 and the Grumman F-14, and of the Soviet Union's MiG-fighter, the Sukhoi Su-24 interdictor and the Tupolev Tu-160 'Blackjack'. Variable wing sweep gives many advantages e.g. in the distances required for take-off, load-carrying ability and in providing a fast low-level penetration role; however, the design is complex and the weight of the aircraft is high. No new variable sweep aircraft have been built since the Tu-160. Needless to say, the pioneering work on variable geometry wings was done in 1949 by the redoubtable Barnes Wallis of bouncing bomb and dam-busting fame.

Gears

Gears are an essential part of rotating machinery of all types, and are of great importance in lifting devices. Early versions of gears were made of wood, with cylindrical pegs for cogs, and they were greased with animal fat. Since a lot of early power generating devices consisted of water wheels of one sort or another, and therefore had a slow speed of rotation, gears had to be used to increase the rotational speed of the business end of the machine to a useful level. Such gears are also used in mills grinding wheat, and an example may still be seen in the Weald and Downland Museum near our home. Gears are a feature of most cars, though our Toyota Prius has an automatic 'planetary' gearbox, which in effect adjusts the gear ratio automatically, so we don't have to worry about changing gear, except to go into reverse.

Most children learn about the use of gears from their bicycles, which have gear boxes of two main types – The Sturmey-Archer, which is incorporated into the rear wheel axle , and the Derailleur, which is a Heath-Robinson-like (but nevertheless effective) structure of cogs around which the bicycle chain is threaded. The Sturmey-Archer Company was founded in 1902 by Henry Sturmey (1857-1930) and James Archer (dates uncertain) under the guidance of Frank Bowden, the primary owner of the Raleigh Bicycle Company. Over the years the number of gears available steadily increased, from the initial three-speed hubs in 1936 to the seven-speed version in 1994, and now even more are often provided. In 2009, Sturmey-Archer reintroduced a three-speed hub, the S3X, which gives ratios of 100/75/62.5, i.e. the top gear is simply direct drive.

The word 'Derailleur' is taken from the French '*derailleur*', derived from the derailment of a train, presumably because when changing gear the chain jumps from one cog to another, in a manner resembling the derailment of a train as it leaves its proper guiding rails. In 1937 the Derailleur system was introduced to the Tour de France. Previously, riders would have to dismount to change their wheels from a downhill to an uphill mode. Derailleur mechanisms may be positioned either on the front or rear wheels. Apart from their more conventional use, they are also employed on mountain bikes, though why anyone would want to cycle up a mountain is beyond me! There are many varieties of Derailleur systems, and their design involves numerous considerations. They have always seemed to me to be very vulnerable to damage, such as when falling off one's bike and hitting the ground. Gear components are variously made of aluminium alloy, steel, plastic or carbon fibre composite, and require moderate lubrication. In most designs the rider has to be actively pedalling at the time of gear change, but versions have been developed in which the rider can be free-wheeling at the time of gear change. There are also electronic versions, in which the gear change is actuated by switches rather than by using manual control levers.

There is currently much encouragement of cycling as a means of transport, since virtually no 'green-house gases' are emitted. Thus towns and cities are increasingly installing cycle lanes, which are meant to be safer for cyclists than simply using the main part of the road. Admittedly, there are still problems to be solved at complex road junctions. Since there is considerable worry about Man's presently largely sedentary life (either sitting at home, or in a train, car or aircraft), cycling clearly provides a useful form of exercise – and it is more interesting and enjoyable than doing twenty press-ups each morning!

Ropes

Ropes are a very ancient tool for making ascents and descents, as well as being used in early ladder-making. Ropes are made of plies, yarns or strands twisted or braided together to increase their strength. They are also used for dragging and pulling. The common fibres used in rope-making are hemp, linen, cotton, coir, jute, straw and sisal.

Coir is a natural fibre extracted from coconuts, the coir fibres being found between the hard internal shell and the outer coat. Mature brown coir fibres have more lignin and less cellulose than other fibres such as flax and cotton. White coir is obtained from unripe coconuts and is smoother and finer, but weaker than brown coir and is used to make mats or ropes. Coir is fairly waterproof and is resistant to damage by salt water. Coconuts themselves are the seeds of the palm *Cocos nucifera*. They can either be harvested after they have fallen to the ground, or climbers may be used and in the best regions may, in this way, pick about 25 trees per day; if armed with a knife attached to a pole, the picker may even harvest up to 250 trees per day. The next stage in the production of coir is the separation of the fibrous layer of the fruit from the hard shell. Manually, this is achieved by smashing the coconut down on to a hard spike to split it, and an expert 'de-husker' can deal with about 2,000 coconuts per day. Machines are available which can also process 2,000 fruit daily.

Brown coir is used in floor mats and doormats, brushes and mattresses (I guess such mattresses are pretty uncomfortable by modern standards). Some brown coir is made into twine and much brown coir is used, after spraying with rubber latex, for upholstery in the automobile manufacturing industry. Brown coir is also used for insulation and padding. White coir has its major use in rope manufacturing, and is also used for making fishing nets, since it is very resistant to saltwater. Since many societies exist largely on fish, the importance of white coir to the well-being of the world population can hardly be over-estimated. India, particularly the state of Kerala, produces over a fifth of the total world supply of white coir, and Sri Lanka produces between 30 and 40% of the total world brown coir output.

Hemp
This is a commonly used name for varieties of the *Cannabis* plant, and apart from its use in the manufacture of fibres of various types, *Cannabis* is also a source of tetrahydrocannabinol, which has major psychoactive properties. The plant's main mechanical (as opposed to psychoactive) use is in the manufacture of paper, textiles, clothing, biodegradable plastics, insulation, health foods and bio-fuel. Nearly half of the weight of hemp seeds consists of edible oils such as the essential fatty acids (i.e. those fatty acids which the body cannot synthesize for itself, such as linoleic, linolenic, eicosapentenoic (EPA) and stearidonic acids – these are fatty acids with long carbon chains). These essential fatty acids ('EFA') are said to be of the omega-3 class, meaning that there are double bond-linked carbon atoms at the third carbon atom from the non-carboxylic end of the molecule. There are many aspects of their 'essentiality'. A few of these are:-

1) They may cause low density lipoproteins (LDL) to change from small to larger, less atherogenic particles, and thus tend to prevent heart attacks and strokes.
2) They may have mild anti-hypertensive effects.
3) They may reduce levels of blood triglycerides, which are also moderately atherogenic.

4) They may reduce the risk of intravascular clotting, possibly by increasing the breakdown of fibrin, a major component of clots.
5) There is evidence that rheumatoid arthritis patients may reduce their pain levels by consumption of EFA.
6) There is some evidence that in patients with dementia, EFA may reduce the rate of cognitive decline.
7) Unbacked by much evidence, EFA are used for the treatment of children with attention deficit hyperactivity disease (ADHD), and autism.
8) EPA supplements may be useful in the treatment of depression.

EFA's do, however, have their dangers. One of the urges of mankind is to take larger and larger doses of compounds or drugs whose effects at normal levels are clearly beneficial. In 2000 the US Food and Drugs Administration warned that consumption of EFA and EPA in excess of 3 grams per day may result in a bleeding tendency, haemorraghic stroke, damage from free radical production during oxidation, increased levels of LDL or *apoproteins associated with LDL in diabetics and hyperlipidaemic subjects, and reduced glycaemic control in diabetics.

Hemp and lime have been used together to construct bricks, but require extra support in the form of brick, wood or steel frames. Hemp has been mixed with fibreglass, kenaf and flax to make composite panels for automobiles. (Kenaf is a fibre obtained from a particular hibiscus plant – *Hibiscus cannabinus*). Of obvious major importance for the advancement of knowledge was the use of hemp in the Chinese Western Han Dynasty, about 2,200 years ago, to make paper. In St. Petersburg, Russia, a paper mill opened in 1818 using hemp, and its output was used in the production of bank notes, postage stamps and various watermarked papers. Hemp has also been used to remove impurities from sewage effluent such as unwanted chemicals, and to crowd out weeds from soil, thus enabling farmers to avoid the use of herbicides. The oil in hemp seeds has been used for the production of biofuels (e.g. biodiesel). China remains the leading producer of hemp, but a large number of other countries also cultivate hemp, including the UK, Canada and Australia.

Ropes for lifting
Though simply hauling on a rope is the simplest method of lifting objects, it is hard work and the process may be enormously eased by pulley systems. A set of pulleys

may be put together in various ways to provide what is the equivalent of gearing in a bicycle or motor car. One of these systems is the *block and tackle*, which is assembled so that one pulley, called 'the block', is attached to a fixed mounting point, and the other is attached to the load to be raised. The mechanical advantage of the block and tackle is equal to the number of parts of the rope that support the block. In a *rope and pulley* system, one end of the rope is fixed to the place to which the load is to be lifted. The rope then descends and passes through a pulley which is attached to the load and then ascends to the destination of the load (weight W), where it passes round another pulley to the hand of the person doing the job. Multiple pulleys may be included to increase the mechanical advantage. If there are n parts of the rope in the system, then the tension in each part of the rope is W/n, which is the input force which has to be exerted to lift the load. The system thus reduces the necessary input force by the factor n. King George III claimed that he had ridden a horse in the Goodwood Cup, a claim whose truth is made improbable by the fact that he was so grossly obese that he was incapable of mounting a horse without being lifted into the saddle with a complicated mechanism involving cranks, winches and rollers. I have looked around Goodwood, which is very near our home, but can find no evidence of such machinery!

Rope ladders

These have the great advantage over ordinary ladders that they can be rolled up and stored away in a relatively small space, and they are also much lighter. Their rungs may be either rigid or flexible, but, being less stable, these ladders are more difficult to use than rigid ladders.

Mountain climbing

This is an occupation to which I am not personally addicted, believing that the more '*firma*' the less '*terra*'. Nevertheless, it seems to be a favourite occupation of those who wish to get an adrenaline 'high' by risking their lives. When the climb is short, the sport is often known as 'rock-climbing'. Barbara and I have spent many holidays in Wharfedale, a few miles north of Leeds in Yorkshire. We have either hired a cottage in a nearby village, such as Kettlewell, or stayed in one of the pubs, either the Tennant Arms in Kilnsey, or the Bluebell in Kettlewell itself or in a cottage in Conistone nearby. Just above the hamlet of Kilnsey is Kilnsey Crag, and there are invariably a few intrepid souls who appear to be pursuing an early death by attempting to climb the Crag, equipped with ropes and crampons. Oddly, we have never heard of anyone actually killing themselves here! In Scotland, we have often stayed at the King's House Hotel near the head of Glencoe, where, from the lounge you can see people attempting to ascend the Buchaille Etive Mor. Friendlier is the habit of local deer walking past the picture window and admiring you as you have dinner. To get some really fierce mountain climbing, the nearest are in the Swiss, French and Italian Alps. The biggest challenge here is the Matterhorn (height 14,690 feet), the first ascent of which was made in 1860 by Edward Whymper, a twenty-year-old English artist, and the enterprise developed into a competition between him and Jean-Antoine Carrel, an Italian mountain guide, who was determined to succeed first for patriotic reasons. One of Carrel's associates, Felice Giordano, had made extensive sketches of the mountain, and declared that it would be necessary to cut steps in the rock over a height of about 100 feet. Whymper had made previous attempts on the mountain, and in 1865 tried a new way, on the east face, but was thwarted by an avalanche of stones, which caused his guides to refuse to make any more attempts by that route. However, another young Englishman, Lord Francis Douglas, turned up, with similar ambitions, and he and Whymper decided to join forces, together with another climber, Charles Hudson, and his friend, Douglas Hadow. They were also accompanied for part of the climb by two experienced guides, Peter Taugwalder and his son, who were acting as porters, and Michel Croz. After they has ascended to 14,000 feet, they rested, and Whymper noted that Hadow, the least experienced climber, required continual assistance, About 200 feet below the summit, Whymper and Croz went on and reached the summit relatively easily. They saw Carrel and his party about 200 metres below.

After an hour they decided to descend, with Croz in the lead, then Hadow, Hudson and Douglas, the Taugwalders bringing up the rear. When they were barely an hour into the descent, Hadow slipped and fell on to Croz, and they both then fell further, pulling down Hudson and Douglas. Whymper and Taugwalder grabbed onto the rocks, but they were unable to stop the fall of their companions, who fell to their deaths over the precipice on to the Matterhorn Glacier, 4,000 feet below. Whymper and Taugwalder (senior) finally reached Zermatt the following morning. A day later, Whymper and Taugwalder discovered the bodies of Croz, Hadow and Hudson, but that of Douglas was never found. There now began an unpleasant controversy. It was noted that those who fell had been tied together by an old, rather weak rope. Whymper was accused of cutting the rope between him and Douglas to save his own life. Taugwalder was tried and acquitted by a court set up by the government of the canton of Valais, but some guides persisted in insisting that Whymper had cut the rope to save himself.

Barbara and I visited Zermatt a few years ago, and walked up to the foot of the mountain, where there is a restaurant at which we lunched, and I painted a picture in water-colour and gouache of the mountain, the summit of which was partly enveloped in cloud, though it cleared enough for me to be able to finish the painting (Plate 3 (b))

Mountaineering in the Himalayas presents a different order of magnitude of problem to Alpine climbing, because, in general, the mountains are about twice as high as in the Alps, and the problems of maintaining blood oxygenation therefore loom large. Altitude sickness (otherwise known as acute mountain sickness) is a pathological effect of high altitude in humans, caused by exposure to low partial pressures of oxygen in the inhaled air. The features one often sees at heights of above 18,000 feet are marked hypoxaemia (low oxygen in the blood) and hypocapnia (low partial pressure of carbon dioxide in the blood due to compensatory over-breathing, the latter causes the blood *pH to rise (become more alkaline) and no permanent human habitation is possible at altitudes greater than about 19,000 feet. Symptoms of mountain sickness include headache, fatigue, vertigo and sleep disturbance. The most dangerous complications are pulmonary oedema (fluid in the lungs) and cerebral oedema (swelling of the brain), The body's response to high altitude includes release of the hormone erythropoietin from the glomeruli of the kidneys, which results in an increase in haemoglobin formatin, the number of red cells per unit volume of blood (polycythaemia) and in the red cell concentration of 2,3-bisphosphoglycerate, a compound which causes a rightwards shift of the*haemoglobin/oxygen dissociation curve (thereby facilitating release of oxygen from red cells more readily). The best way of preventing altitude sickness is acclimatization by making a very gradual ascent, thus allowing the above compensatory mechanisms time to develop. The condition may, however, be partly prevented by the drug acetazolamide, a carbonic anhydrase inhibitor, by using supplemental oxygen, or, (better still), by giving up the climb and descending!

As every schoolchild knows, Everest is the world's highest mountain (height 29,029 feet above sea level) and may be still rising as a result of gradual movement of tectonic plates. In 1921, a reconnaissance expedition was mounted, to survey the possibility of climbing it; on it, George Mallory and colleagues tried to discover whether a route could be found from the north side. In 1922, a British expedition mounted a full-scale attempt on the summit. A day later, two other members of the group, George Finch and Geoffrey Bruce climbed the North ridge (27,300 feet), using oxygen for the first time. During this expedition, several *Sherpas were killed by an avalanche. In 1924, Mallory and Irvine left their high camp at 26,900 feet, using oxygen. This was the last time the pair was seen alive, and whether one or both of them reached the summit is an unsolved question. In 1934, a British eccentric, Maurice Wilson, set out to climb Everest alone and his body was found by a later expedition. In 1935, Eric Shipton led a reconnaissance group in preparation for an attack on the summit the following year and noted an ice-filled basin at the foot of the main summit known as the Western Cwm. This expedition was notable for the first involvement of the Sherpa Tensing Norgay, who eventually reached the summit with Edmund Hillary in 1952. Sir John Hunt, the leader of the expedition, sent his first pair of climbers, Charles Evans and Tom Bourdillon, who got to within 300 feet of the summit, but had to descend because of problems with the oxygen equipment and lack of daylight time. Two days later, Hunt sent in his second pair of climbers, the New Zealander Edmund Hillary and, again, Tensing Norgay, who finally reached the summit on 23rd May, 1953. There was intense (and rather stupid) speculation on which of the pair first set foot on the summit, and conveniently, the news reached London on the morning of the coronation of Queen Elizabeth II. A few years later, the controversy about the 'first-footing' was closed by Tensing, who revealed that it was Hillary. In 1978, the extraordinary feat of climbing Everest without the use of oxygen was achieved by Reinhold Messner and Peter Habeler. The physiological effects of this are described by Messner as feeling that "he no longer belonged to himself and, to my eyesight, I am nothing more than a single narrow thing, floating over the mists and summits". These two climbers had previously practised climbing without oxygen on the North Face of the Eiger (the Eiger Nordwand), and the Matterhorn.

Near to Everest are the several other great Himalayan peaks Makalu, Lhotse and Nuptse. Makalu (27,825 feet) is the fifth highest mountain in the world. It was first climbed in 1955 by Lionel Terray and Jean Couzy. Lhotze (27,605 feet) was first climbed in 1956 by the Swiss team of Ernst Reiss and Fritz Luchsinger; it has since been climbed by 371 climbers and 20 have died in the attempt. Nuptse (25,790 feet) was first climbed in 1961 by Dennis Davis and Sherpa Tashi. One of the early climbers of Nuptse was Jim Swallow, a medical colleague of ours at The London Hospital, and the time off that he took to make the climb became known locally as

'Jim Swallow's nuptial flight'. Annapurna (26,545 feet) is in a section of the Himalayas in north-central Nepal, and it has the reputation of being extremely dangerous, as judged by the 'fatality-to-summit ratio of 38% which is the highest of any peak above 8,000 metres (about 27,400 feet). It was first climbed in 1970 by Don Whillans and Dougal Haston.

The Sherpa has obviously played a large part in Himalayan mountaineering, acting as climbers, guides and porters. Most Sherpas live in the East of Nepal. The Sherpa language is a Tibeto-Burman tongue, which is unintelligible to Tibetans who live in the capital, Lhasa. It is claimed that Sherpas live at those high altitudes because of genetic adaptations. Some of these include unique oxygen-haemoglobin-binding enzymes and high *nitric oxide (NO) production. The later causes a left-shift of the haemoglobin-oxygen dissociation curve, implying that the haemoglobin is better able to hold on to what little oxygen is in the atmosphere at those heights. The religion of Sherpas is Buddhism, but they also believe in numerous deities and demons that inhabit caves, mountains and forests. The monastery is an important feature of the Sherpa religion, and there are communities of lamas, monks and nuns who take a vow of celibacy and spend their lives seeking for truth and enlightenment. In 2012, sixteen-year-old Nima Chhamzi Sherpa became the youngest woman to climb Everest.

Rockets

Rockets are, at present, the only method of escaping both the Earth's atmospheres and its gravitational field. But before considering that sort of enterprise, let us first start at the most mundane level, namely with fireworks, traditionally let off in the UK on 5th November each year to commemorate the unsuccessful attempt by Guy Fawkes and colleagues to blow up the Houses of Parliament. Fawkes (1570-1606) was born and educated in York. He later converted to Roman Catholicism and fought in the Eighty Years War on the side of Catholic Spain against Protestant Dutch reformers. He unsuccessfully sought support for a Catholic rebellion in England, but later he was introduced to Robert Catesby, who planned to assassinate King James I and restore a Catholic monarch to the throne. The first meeting of the plotters (Fawkes, Thomas Wintour, Robert Catesby, an exiled Welsh spy named Hugh Owen and Sir William Stanley) took place at an inn known as 'The Duck and Drake', near The Strand. The plotters obtained a lease on a chamber under the House of Lords and Fawkes was placed in charge of the 20 barrels of gunpowder stockpiled there. However, someone whose identity is unknown, betrayed them, and Fawkes was captured by the guards, tortured in several ways, including the use of the rack, until eventually he revealed details of the plot. Immediately before he was about to be hung, he jumped from the scaffold and broke his neck, thereby escaping being hung, drawn and quartered!

The 5th of November is variously known as Guy Fawkes Night, Plot Night, and Bonfire Night. Apart from the firework displays, effigies, usually of Guy Fawkes or the Pope, are also burnt, and on 5th November in England to this day the sky at night is lit up by displays of pyrotechnics. These have also also used to celebrate notable events, such as the end of World War II, the Queen's coronation and various jubilees, both personal and national. The bonfires are an excuse for everyone to communally burn their rubbish, and in our village of East Dean the fuel for the bonfire grows into an enormous mountain on the village green. Traditionally, in East Dean, bonfire night is accompanied by the consumption of soup, hot dogs, well-garnished with mustard, and cooked and sold by the roadside. There are two downsides to this whole caper; firstly, the village green is ruined by the bonfire and takes some months to regrow, and secondly, the following morning, our garden is usually bestrewn with burnt-out rocket cases together with the sticks which are used to support them in milk bottles so that the touch-paper may be easily lit.

Military uses of rockets

The bazooka is an anti-tank weapon wielded by a single soldier, also referred to as 'The Stovepipe'. It has a solid-fuel rocket motor for propulsion and this enables its missile to be delivered against armoured vehicles, machine gun posts and other fortified bunkers beyond the range of a hand-thrown grenade. Its missiles can penetrate up to 2.4 inches of armour-plating. The name arises from a musical instrument called a 'bazooka', invented in the 1930's by a US comedian, Bob Burns. After the Germans captured some bazookas on the Eastern Front and in the North African campaign, they made their own version, the *Roketenpanzerbüche,* or *Panzerschreck,* i.e. the 'Tank Terror'.

The Patriot Missile is a surface-to-air weapon launched from vehicles and is guided by a computer and radar to its targets, which have usually been incoming ballistic missiles. The missile only has to explode in the vicinity of the target to destroy it with the shrapnel released by its explosion. The Patriot missile is 5.8 metres long and weighs about 900 kilograms. The launching vehicles have several launching tubes, so that a number of missiles may be fired simultaneously. They have been used in the Persian Gulf War (Operation Desert Storm, 1991) to shoot down incoming Iraqi Scud missiles. In February, 1991, an Iraqi Scud hit the US barracks at Dhahran, Saudi Arabia, killing 28 soldiers. The failure to intercept this missile was found to be due to a software error in the system's clock; the apparatus had successfully detected the incoming Scud, but because of the error, had looked for it in the wrong part of the sky. In Operation Iraqi Freedom (2003), modified Patriot missiles achieved a high success rate in the interception of Al-Samoud 2 and Ababil-100 tactical ballistic missiles. However, there were downsides – a Royal Air Force Tornado was destroyed in error by a Patriot missile, resulting in the death of the two crew members, and a United States Air Force Fighting Falcon jet fired at a Patriot missile battery, after mistaking the Patriot for an Iraqi surface-to-air missile. Also in 2003, two Patriot missiles shot down a United States Navy Hornet fighter, killing the pilot.

The V-1 flying bomb, also known as the Buzz Bomb or Doodlebug was developed by the Germans at Peenemunde Army Research Station by the Luftwaffe during World War II. It was one of the *Vergeltungswaffen* series of weapons designed to terrorize London and the East Coast. It was really one of the few desperate last throws of the Germans in an attempt to seize the initiative again immediately after the D-Day landings of 6th June. At the peak of its deployment, more than one hundred per day were launched from sites in the Pas de Calais and the Dutch coast. The total number of launches was over 9,500, decreasing in number as the launch sites were overrun by the Allies. The last V-1 was in fact directed at the port of Antwerp. These missiles were fuelled by high octane gasoline, but acetylene was used for starting the launch. They were guided by a gyroscopic system but were very inaccurate and only suitable for general area bombing. An *odometer driven by an anemometer in the nose cone determined when the target area had been reached, and the missile was then put into a steep dive. This dive was caused by the cessation of fuel flow, so that the inhabitants of the target area soon learnt that when the roar of the rocket ceased, they should take immediate cover. Some of these missiles were intended to be piggy-backed on the top of the fuselages of Arado Ar 34 bombers, but these were never used. They were also employed as reserve fuel tanks towed behind the Messerschmidt Me 262 jet fighter. Defences against the V-1 before D-day consisted of light anti-aircraft guns, but the V-1 cruising height was just above the range of these guns. The defence situation was improved by the development of proximity fuses, so that the missile did not actually have to be hit, and also by radar guidance systems for anti-missile missiles. The speed of the V-1 in horizontal flight was about 350 mph, and the only aircraft at that time able to keep up with them was the Hawker Tempest, whose numbers were gradually built up and they were joined by P-51 Mustangs and Supermarine Spitfires (Mark XIV). V-1s were also counteracted by leaking inaccurate information as to which structures they had hit to German agents in Britain who had been 'turned'. By September, 1944, 4,261 V-1s had been destroyed by fighters, barrage balloons and anti-aircraft fire. The main benefit to the Germans resulting from the use of the V-1 was that the Allies were forced to invest heavily in anti-V-1 defences, and the bombers used to destroy the Axis launching sites had to be diverted from other more important targets.

The next German rocket weapon was the V-2 (*Vergeltungswaffe* – retaliation weapon). This is much more significant historically, because it was the fore-runner of all later rockets developed both for both the United States and the Soviet space programmes. It was the world's first long-range combat missile and the first human machine to break free of the Earth's gravitational field and enter space. It was also the rocket which made the name of Wernher von Braun. The V-2s were constructed at the Mittelwerk site by prisoners from Mittelbau-Dora, a concentration camp in which about 20,000 prisoners died during the World War II. This rocket used a 75% ethanol/water mixture for fuel and liquid oxygen to provide the oxidiser. It had fins as an essential part of the guidance system. The first V-2s had engine cut-off systems which were triggered when the target was about to be reached, but later V-2s used guide beams. The V-2s were first launched from bunkers or from fixed pads, but these were eventually abandoned for mobile launchers, which were obviously less vulnerable to destruction by the Allies. The British Government initially tried to conceal the fact that a new weapon was being used by the Germans, but in November 1944, Winston Churchill informed Parliament that England had been under attack from these rockets for the past few weeks. Over 3,500 were launched at targets in Belgium, the UK, France, The Netherlands and eventually at targets in Germany itself, for example, at the bridge over the Rhine at Remagen. Over 2,700 Londoners were killed by V-2s and a further 6,523 injured. One of the counter-measures used by the British was to spread dis-information suggesting that the rockets were overshooting their London targets by 10-20 miles; this was successful, and the re-programming of the missiles by the Germans caused them to land in less heavily populated areas in Kent. Other countermeasures, such as fighters and anti-aircraft fire were useless, so that the only other effective approach was to destroy the launch sites. Ultimately, the Allied advances in Europe after D-Day forced the mobile launchers so far back that their missiles could no longer reach their targets.

Wernher Magnus, Freiherr von Braun (1912-1977) was a German rocket scientist and aerospace engineer who is generally known as the father of rocket science. He was born in Wirsitz, in the Province of Posen, of an aristocratic family. He was, according to his mother, the family genealogist, ultimately descended from Charlemagne (742-814 AD)! His career began at the age of twelve when he caused major disruption in a crowded street by detonating a toy wagon to which he had attached some fireworks. He was a talented musician, playing Beethoven and Bach from memory, and took lessons from the composer and pianist Paul Hindemith. He attended a lecture given by Auguste Piccard, and blithely announced to him that he intended to travel to the Moon sometime. One of his experimental rockets worked perfectly, except that it landed on the wrong planet! Von Braun joined the Nazi Party in 1937 and at that time was mildly enthusiastic about Hitler, who by then was tearing up the end of World War I Versailles Treaty, reoccupying the Rhineland and absorbing Austria and the Sudetenland into the Reich. He joined the SS at the behest of Himmler. His rockets were powered by alcohol and liquid oxygen, as opposed to those built by the firm of Hellmuth Walter at Kiel to provide a rocket engine for the Heinkel 112, which used hydrogen peroxide and calcium permanganate as the catalyst. However, subsequent flights with the Heinkel 112 used the Walter engine rather than von Braun's, since it was more reliable and less likely to kill the test pilot. Von Braun was later heavily criticised for using slave labour for the rocket programme, but denied any knowledge of the concentration camp conditions under which the workers had been held. When asked why he had not protested against those conditions, he replied that he would have been shot if he had. In 1943, von Braun was arrested on a charge of being a Communist sympathiser and attempting to sabotage the V-2 programme. He was taken to a Gestapo cell in Stettin, but Albert Speer, Hitler's munitions chief, persuaded his master to re-instate von Braun so that the V-2 programme could proceed.

In the Spring of 1945 the Soviet Army had advanced to within 160 km of Peenemünde, but von Braun and his colleagues were determined to surrender to the Americans, because of the Soviet reputation for cruelty. In May, von Braun and his brother Magnus surrendered to an American private on a bicycle, and made an announcement to the press that they wanted to surrender to a people guided by the

Bible, since only in that way could the safety of the world be assured He was then recruited into the US Army and transferred to America! Here his period of service to the Nazis was expunged from the record, so that he could serve US interests without his background being questioned. He was housed at Fort Bliss, an army installation just north of El Paso, and later transferred to the White Sands proving ground in New Mexico, where the use of rockets for military and research applications was being studied. In 1950, at the start of the Korean War, he was transferred to Huntsville, Alabama, and he lived there for the next 20 years. His ambition was to set a man on the surface of the Moon, and this was realised on 16th July, 1969 with the success of the Apollo 11 mission, just a few days before his death.

However, all was not one continuous stream of exciting achievement in the US rocket programme. In 1986 the Space Shuttle '*Challenger*' disintegrated 73 seconds into a mission, leading to the deaths of its seven crew members. A rubber O-ring seal in one of its booster rockets had failed, allowing pressurised hot gas to escape and impinge on the adjacent solid fuel booster. The President, Ronald Reagan, instituted a special Commission to investigate the disaster. It turned out that the propensity for failure in the O-rings had been known for several years, but no-one had acted to correct the fault. One of the problems was that the launch took place in the early morning, when ambient temperatures were low (far lower than had been the case in earlier launches), and the rubber of the O-ring was therefore rather stiff; this impeded the sealing qualities of the O-rings. One of the Commission's members was the physicist Richard Feynman, who, during a televised hearing, demonstrated how the O-rings became less resilient (and therefore more likely to fail at icy temperatures) by immersing a sample of the O-ring material in a glass of iced water. He showed that the estimates of reliability made by NASA were very unrealistic, differing by up to a thousand-fold from those of proper working engineers. He concluded that "reality must take precedence over public relations, since Nature cannot be fooled". One of the crew members was Christa McAuliffe, who would have been the first teacher in space. The disaster prompted that consummate actor and orator, Ronald Regan, to quote in his speech to the nation a poem by John Gillespie Magee:-

"we will never forget them, nor the last time we saw them, this morning, as they prepared for their journey and waved goodbye and slipped the surly bonds of Earth to touch the face of God"

Going down
More or less everything we have so far considered has been about how things may be approached in an upward direction, but we must now consider Man's efforts at descents.

Sea diving
Diving by individuals has significant major hazards. Apart from the incidental danger of being attacked by sharks that may be lurking nearby, and failure of one's diving equipment (e.g. oxygen supply), the main potential problem is decompression sickness (DCS). This occurs on ascent, and arises from gases (mainly nitrogen) that have dissolved in the body fluids at depth coming out of solution and lodging in the blood vessels of crucial organs. The effects of decompression sickness vary from joint pains, rashes, headaches, vertigo, epilepsy and visual abnormalities to paralysis and death. Joint pains are often referred to as 'the bends'. Other terms used are 'the chokes' for breathing problems and 'the staggers' for neurological problems. Though DCS is usually associated with diving, it may also occur in caisson workers, those who fly in unpressurised aircraft, and in spacewalkers. The condition may be prevented in divers by slow ascents, with frequent stops, allowing the dissolved gaes to emerge from solution and escape. The risk of DCS is increased the longer the dive time, and the deeper the dive. In addition, the shorter the interval between dives, the less time is available for the dissolved gases to be off-loaded safely through the lungs. Rapid ascents can cause permanent bone damage (dysbaric *osteonecrosis). Passenger airliners are pressurised to an altitude equivalent to 7,900 ft to prevent these problems. To avoid DCS, professional divers limit their rate of ascent to about 33ft per minute, or use a schedule derived from published decompression tables. They also have to make sure that the dissolved gases are fully eliminated before diving again, and, again, published tables give guidance as to the time required, which may be up to 18 hours. Breathing pure oxygen for a period before diving reduces the nitrogen dissolved in body fluids. Once it has developed, DCS is treated with 100% oxygen delivered in a high-pressure chamber ('hyperbaric oxygen).

Notable outbreaks of DCS occurred during the construction of the Brooklyn Bridge in 1873 and the Eads Bridge in St. Louis, Missouri in 1871. The outbreak in the Brooklyn Bridge project was termed 'caisson disease', and afflicted, amongst others, the chief engineer, Washington Roebling. During this project, the disease became known as 'The Grecian Bends', because the sufferers characteristically arch their backs, in a posture resembling that taken in a women's dance step known as the Grecian Bend. In Jules Verne's science fiction thriller '*A Journey to the Centre of the Earth*', published in 1864, a German Professor, Otto Lidenbrock, believes that there are volcanic tubes penetrating to the centre of the Earth and he and his companions descend into the crater of the Icelandic volcano, Snaefellsjökull, encountering many prehistoric animals and other natural hazards, before emerging through the crater of the volcano Stromboli in southern Italy.

The shafts of coal mines may be enormously deep, and may branch out into separate galleries at many levels, some of these going under the sea. There has been some dispute about which is the deepest shaft in the UK. *Hansard reports that in 1934 the deepest mine was the Parsonage Pit (3,850 feet); apparently this depth has now been increased to over 4,000 feet. At a debate in Parliament on the subject of mining, it was stated that the miners at that depth were complaining of excessive heat, the temperature being recorded as 115 degrees Fahrenheit. Other contenders for the 'deepest pit' accolade are the Clock Face pit near Newcastle, and the Wolstanton No.2 pit in North Staffordshire (?3,750 feet). Paul L. Younger, Director of the Joseph Swan Institute for Energy Research at Newcastle University quotes Aneurin Bevan (who himself came from a mining community in South Wales):-

"This island is made mainly of coal and is surrounded by fish. Only an organising genius could produce a shortage of coal and fish in Great Britain at the same time".

However, this quote may possibly relate to late 19th century Cornwall, when the mining industry collapsed (presumably tin mines, of when there are still many ruins to be seen in Cornwall) and the pilchards vanished.

Aneurin Bevan ('Nye') (1897-1960) was a Welsh Labour Party politician who became Minister of Health from 1945-1951 in the Government of Clement Attlee. He won a scholarship to the Central Labour College in London, where he read economics, politics and history. He was plagued by a severe stammer, but overcame this, it is said, by reading long passages from the works of William Morris, an English textile designer, artist, writer and libertarian socialist associated with the Pre-Raphaelite Brotherhood., who founded the Socialist League in 1884. He joined the South West Wales Miner's Federation and thus became a Trade Union activist. At the 1929 general Election he won the seat of Ebbw Vale for the Labour Party, and made fiery speeches, criticising the Conservative Winston Churchill and the Liberal, David Lloyd George, both of whom, he felt, opposed the working man. He was regarded as a trouble maker by the manager of his employers, the Tredegar Iron and Coal Company, and some excuse was trumped up to sack him, but Bevan took the Company to court, won his case and was re-instated. He was largely responsible for the introduction of the National Health Service, which provided free health care at the point of delivery for everyone, but resigned when the Attlee government proposed to charge patients a fee for spectacles and false teeth. I remember that on the day on which the National Health Service started, my father, a general practitioner in Plymouth, glumly remarked "we're all bloody civil servants now!" There had been previous schemes to ease the burden of payment for health services on the poorer members of society, but these were '*ad hoc*' and local. One of these, in which my father participated, was known as 'The Panel'. For a small regular subscription to the Panel, one could have one's medical fees spread out and somewhat reduced, a sort of primitive insurance scheme. To give Bevan his due, he was a strong critic of the policies of Neville Chamberlain, who negotiated with Adolf Hitler in a desperate attempt to keep Britain out of a second World War, and returned to the UK waving a piece of paper (which was an apparent agreement for avoidance of war), saying that he had achieved "peace for our time". Of course, Hitler had no intention of sticking to the agreement.

During the Second World War, Bevan was critical of the leadership of the British Army, which he felt was 'class bound' and ineffective. After Rommel's defeat of General Auchinleck's army, and, in a motion of censure against the Churchill Government, he declared "The Prime Minister must realise that in this country there is a taunt on everyone's lips that if Rommel had been in the British Army, he would still have been a sergeant". This seems unrealistic, since Erwin Rommel, the so-called 'Desert Fox', was respected by friend and foe alike as a humane and professional officer and his *Afrika Corps* was never accused of war crimes. Orders from Hitler to kill Jewish soldiers were ignored. Rommel was also linked to the conspiracy to assassinate Hitler, but because he was a national hero, Hitler sought to eliminate him quietly. In return for assurances that his family would not be persecuted after his death, he was allowed him to commit suicide with a cyanide pill. He was given a state funeral and it was announced that he had died of old battle wounds.

Winston Churchill (1874-1965)

Churchill was born into an aristocratic family, and most certainly did not suffer from the troubles from inbreeding seen in so many such families who regarded it as necessary for their members to marry one of their own class. He was the grandson of the 7th Duke of Marlborough. His father, Lord Randolph Churchill, had been Chancellor of the Exchequer, and his mother, Jenny Jerome, was an American socialite. In his youth he was an army officer, seeing action in India, The Sudan and in the Second Boer War. He gained a reputation as a successful war correspondent and wrote books about his campaigns, e.g. 'My Early Life'. He became the Member of Parliament for the constituency of Oldham in 1900. Before World War I, he served as President of the Board of Trade, Home Secretary, and First Lord of the Admiralty as part of Asquith's government. He somewhat blotted his copy book on the *Gallipoli (Dardanelles) campaign against the Ottoman Turks, and was demoted from the post of The First Lord of the Admiralty to the sinecure of Chancellor of the Duchy of Lancaster. However, his star began to rise again when Asquith made him Minister of Munitions. He was appointed Chancellor of the Exchequer in 1924 under Stanley Baldwin, and oversaw Britain's ill-conceived return to the Gold Standard, which resulted in unemployment and a miner's strike. During the General Strike of 1926, Churchill argued that either the country would break the General Strike, or the General Strike would break the country. He lost his ministerial office and spent the next few years concentrating on his writing. One of Churchill's great gifts was his ability to write first-class prose, and during this time he produced 'Marlborough – his Life and Times', 'A History of the English-Speaking Peoples', 'Great Contemporaries' and further developed his political views, which curiously seemed to advocate an abandonment of universal suffrage, and a return to a franchise based on property. After World War II, he wrote a history of that event in four volumes, which have now become a classic.

Churchill opposed the Indian independence movement led by Mahatma Gandhi; the latter went on hunger strikes to gain his political ends, knowing that his death would provoke violence on a massive scale. Churchill was quoted as favouring letting him die the next time he went on hunger strike. He declared that "it is alarming and also nauseating to see Mr. Gandhi, a seditious Middle Temple lawyer, now posing as a fakir of a type well-known in the East, striding half-naked up the steps of the Vice-Regal palace to parley on equal terms with the representative of the King-Emperor". Some historians blame Churchill for the high mortality (?3 million people) in the

Bengal famine of 1943, but others maintain that the real cause was the fall of Burma to the Japanese, which cut off India's main source of rice imports. He was remarkably insensitive, asking in a telegram to Lord Wavell, the Viceroy, why, if food was so scarce, Gandhi hadn't yet died.

Churchill's reputation began to be restored when he was one of the first to draw attention to the problem of German rearmament, in defiance of the Versailles Treaty. He stressed the need to rebuild the Royal Air Force, and to create a Ministry of Defence, and urged a renewed role for the League of Nations. However, Churchill was passed over for the post of Ministry of Defence, in favour of the Attorney General, Sir Thomas Inskip. The historian Alan Taylor described that appointment as the most extraordinary since Caligula made his horse a Consul! Churchill's reputation was again damaged in the Abdication Crisis of 1936, when he urged delay in King Edward's stepping down to marry Mrs. Wallis Simpson, implying that unconstitutional pressure was being applied to the King to force him into a hasty decision.

Adolf Eichmann (1906-1962)

Eichmann was one of the major organisers of the Holocaust, and was eventually successfully pursued and captured by Israeli agents. He was born to a Lutheran family in Solingen, Germany, son of a businessman and industrialist. After his mother died, the family moved to Linz, in Austria. He left high school without having any formal graduation, and started working in his father's mining company, and in 1933 he moved back to Germany. On the advice of a family friend, Ernst Kaltenbrunner, he joined the Austrian branch of the Nazi Party, and became a member of the *Schutzstaffel* (SS) (Knights of the Holy See). He was assigned to the administrative staff of Dachau concentration camp, but in 1934 requested transfer to the *Sicherheitsdienst*, the security service of the SS, because he wished to get away from the monotony of training at Dachau. In 1937 he was commissioned as an SS-*Untersturmführer*, equivalent to a second lieutenant.

Also in 1937 he travelled to Israel to assess the possibilities of expulsion of the Jews to that country being a solution to the 'Jewish Problem'. Using false credentials he and an accompanying colleague spent two days in Haifa, but were refused re-entry into Palestine by the British authorities. In 1938 he was sent to Austria to help organise SS security forces after the *Anschluss with Germany. In 1941 Reinhard

Heydrich told Eichmann that all Jews in German-controlled Europe were to be murdered, and ordered him to attend the notorious Wannsee conference as its secretary. This was where the plans for the 'Final Solution to the Jewish Question' were made. Eichmann was given the role of transport administrator, which put him in charge of all the trains transporting Jews to the death camps in Poland. He first made an offer to trade captive Jews to the West in return for trucks and other goods. When this proposal was turned down, he started deporting Hungarian Jews, sending no less than 430,000 to death in the gas chambers. In November, 1944, when 'the writing was on the wall' for Nazi Germany, Heinrich Himmler ordered Jewish extermination to be halted and all evidence of 'The Final Solution' to be destroyed, but Eichmann, who was shocked by Himmler's action, simply carried on against official orders. In late 1944, Eichmann fled from Budapest to avoid being captured by the advancing Russians, returning to Berlin and then again to Austria, where he once more met up with Kaltenbrunner. The latter, however, refused to associate with him since his duties as an extermination administrator had made him a marked man so far as the Allies were concerned.

At the end of the war, Eichmann was captured by US forces, who were not, however, aware of his true identity. In 1946 he escaped, and hid in Altensalzkoth, a small village on Lüneberg Heath. At the beginning of 1950, he went to Italy, where he posed as a refugee. With the help of Bishop Alois Hudal, an Austrian cleric who organised escape routes for Axis criminals, he obtained an International Red Cross humanitarian passport. He boarded a ship bound for Argentina in 1950, and later managed to arrange for his family to join him. For ten years he worked in a succession of jobs in Buenos Aires.

However, he reckoned without the activities of Mossad, Israel's intelligence agency, and those of Simon Wiesenthal, a Jewish Nazi hunter. In 1954, Wiesenthal saw a letter received by an Austrian living in Buenos Aires, saying:-

"I saw that filthy pig, Eichmann. He lives near Buenos Aires and works for a water company".

In 1959, *Mossad received further information to the effect that Eichmann was in Buenos Aires, living under the assumed name, Riccardo Klement, and the Israeli government approved a covert operation to capture him. Mossad managed to blind him temporarily with their car headlights, seized him and took him to a Mossad safe house. He was then drugged to appear drunk, and was smuggled out of Argentina on an El Al plane, eventually beimng landed in Israel on 21st May, 1960. The Israeli Prime Minister David Ben-Gurion then announced Eichmann's capture in the Knesset (Israel's parliament), an announcement that received a standing ovation. After an international dispute over the legality of these operations, Eichmann was eventually brought to trial. Eichmann sat in a dock surrounded by bullet-proof glass to protect him from his victims' families. He did not dispute the facts of the Holocaust, and used the defence so often used in the 1945-1946 Nuremberg trials, insisting that he was only "following orders", either from Hitler himself, or Eichmann's superior officers. After fourteen weeks of testimony, supported by 1,500 documents and 100 prosecution witnesses, many of whom were survivors of concentration camps, he was found guilty and sentenced to death. He was hanged on 31st May, 1962, at a prison in Ramla, Israel. 'Following orders' is a common excuse made for their actions by members of dictatorships, and carries little weight in civilised society

Bridges

The first bridges were made by Nature rather than by Man, and probably consisted of tree trunks conveniently falling across streams. Both trees and bamboo poles were used by early Americans. Sticks, logs and tree branches were lashed together with rope to add substance and strength to the bridge. In the Peloponnese, in Greece, the Arkidako Bridge was built in the 13th century BC to accommodate chariots, and was based on a series of arches; it is still in use. The Romans were major bridge builders, no doubt to facilitate their conquests. Another example is the Alcantara Bridge, over the River Tagus in Spain. The builders used a type of cement (*pozzolana*), consisting of water, lime, sand and volcanic rock. In India, the Mughal administration built bridges, using plaited bamboo and iron, for military and commercial purposes. In the 6th and 7th centuries AD, the Chinese built stone bridges, such as the Zhaozhou Bridge, during the Sui Dynasty and rope bridges were used by the Incas of Peru in the 16th century.

Another famous bridge in Rome was that over the River Tiber, which was defended by the sole efforts of Horatius Cocles from the invading army of Lars Porsena, King of Clusium in the late 6th century BC, who was assisting the deposed last king of Rome, Lucius Tarquinius Superbus. Though gravely wounded, he threw back enemy missiles, and defended the bridge with his sword, a good example of how a superior force may be resisted by mounting the defence on a narrow front, e.g. as did Greece against the Persian army of Xerxes in the Pass of Thermopylae in 480 BC . This episode was immortalised by Thomas Babington Macaulay (1800-1859):-

"Lars Porsena of Clusium, by the Nine Gods he swore
That the great house of Tarquin should suffer wrong no more.
By the Nine Gods he swore it, and named a trysting day,
And bade his messengers ride forth,
East and West and South and North,
To summon his array."
(etc., etc.)

In the UK, the Iron Bridge crossing the River Severn in Shropshire was the first in the world to be made of cast iron (previously the only way to cross the Severn Gorge had been by ferry). This had been an expensive material, which deterred its construction for some time, but a new blast furnace, set up nearby, lowered the cost and work was able to start in 1779, the bridge being opened on New Year's Day, 1781. Since 1934 it has been closed to vehicular traffic, but Barbara and I have walked across it and enjoyed the spectacular views over the Gorge. It has now been designated a Scheduled Ancient Monument, and the bridge and the adjacent area form the UNESCO Ironbridge Gorge World Heritage Site.

In my childhood, our family made frequent visits to beauty spots in the Dartmoor area, which is just north of Plymouth. Amongst the places we visited were very ancient bridges with nooks in the span into which pedestrians could retreat to allow vehicles such as wagons and horse-drawn carts to pass. There were also clapper bridges; these are constructed from large flat slabs of granite or schist, supported on stone piers built at intervals across rivers. Most clapper bridges were erected in mediaeval times, or later. The word 'clapper', according to the wardens of the Dartmoor National Park, is derived from the Anglo-Saxon word *cleaca*, which means 'bridging the stepping stones'. If this is true, Anglo-Saxon must be a remarkable language to be able to express a fairly complex concept in such a short word! One of the best examples of a clapper bridge may be seen at Postbridge on Dartmoor (Plate 3(c)); its slabs are 13ft long and 6ft 6 inches wide, and weigh over eight tons each. This bridge first appeared in a mediaeval description written in 1380AD, and was built to facilitate the transport of tin from the Dartmoor mines to the stannary town of Tavistock by pack horse. A 'stannary' is a term either applied to a tin mine or to a town engaged in refining and assaying the tin. One of a stannary town's principal roles was the making of tin coinage, which was passed to the Duchy of Cornwall or to the Crown. The village of Plympton, just east of Plymouth, became a stannary town in 1328 after the Sheriff of Devon was persuaded that it was near to the sea and therefore provided good access for tin merchants. During the 2nd World War, my mother, sister and myself made frequent visits to Plympton to pick up goods on the 'black market', to supplement what we got from our ration books (everyone was doing this sort of thing to survive!). Plympton was the birthplace of Sir Joshua Reynolds (1723-1792), the distinguished portrait painter who became the first President of the Royal Academy. In 1749 he was a passenger on HMS *Centurion* on a trip to the countries of the Mediterranean, where he spent some years studying the works of the Old Masters. On this trip he acquired an ear infection which left him partially deaf, and he is often pictured carrying an ear trumpet. He established an atelier in Leicester Square, London, to which the great and the good came to have their portraits painted, and would deal with five or six sitters a day. The clothing of his subjects was usually painted by one of his pupils or assistants. He was especially good at painting children, not an easy task, since they are less differentiated in their features than in adult life, and are not very good at sitting still! He had many friends amongst the London intelligentsia, including Samuel Johnson (of English Dictionary fame), Oliver Goldsmith (playwright), Edmund Burke (statesman and orator) and David Garrick (actor, playwright and theatre manager).

I was at school at Clifton College Bristol, which is about a mile from the Avon Gorge; this school had some of its sports grounds at a place named Beggar's Bush, on the far side of the Gorge, and I made frequent trips across the Clifton Suspension Bridge to indulge (somewhat reluctantly) in rugby football and cricket. My revenge was to develop a reputation for having the nastiest hand-off in the School. For the uninitiated, a hand-off is a legitimate means of warding off someone in the other team who is attempting to tackle you. The hand-off must be made with the flat of the hand, but every now and again tempers frayed, and the hand-off turned into a punch. If this was spotted by the referee, one was liable to be sent off the field in disgrace. Clifton Suspension Bridge is a favourite site for people with suicidal intent to jump off. On one occasion, in the age of crinolines, one lady trying to commit *felo de se* was thwarted because the crinoline acted as a parachute and she landed unharmed! It has been the venue for a number of notable events, e.g. the first bungee jump, the hand-over of the Olympic Torch relay in 2012, and the last ever Concorde flight in 2003 passed over it, presumably either on its way to Filton airport near Bristol or to Heathrow, where a model of the plane stood at the entrance to the underground tunnel into the airport for many years. Clifton Suspension Bridge has a span of over 700 feet, and is about 200 feet above the River Avon. This iconic suspension bridge was designed by the celebrated Isambard Kingdom Brunel (1806-1859). He was an English civil and mechanical engineer. He worked for several years on the Rotherhithe Tunnel under the River Thames (Barbara and I used this for many years to travel to work at The London Hospital, Whitechapel from our home in Beckenham). The construction of the Rotherhithe Tunnel claimed the lives of several workers, when it became flooded from the Thames above.

At the time of its construction the Clifton Suspension Bridge had the longest span of any bridge in the world. To get the commission, Brunel had to submit a number of designs to a committee chaired by Thomas Telford, who rejected all of them, proposing a design of his own instead! There was an outcry from the public and eventually a new competition was held, which was won by Brunel. Work was not completed until 1864, five years after Brunel's death. Approximately four million vehicles cross this bridge every year. Brunel was also involved in the construction of railways, and for these projects he designed many other bridges, including the Royal Albert Bridge, spanning the River Tamar at Saltash near Plymouth. It should be noted that this is purely a railway bridge, and for many years when we drove into Cornwall from my parents' home in Plymouth, we had to make an enormous detour over Dartmoor to the village of Gunnislake, where the river was narrow enough to boast a bridge of considerable antiquity. Whilst this was an attractively scenic route, it added several hours to the journey. The Royal Albert Bridge was designed by Brunel in 1855, after Parliament had rejected a plan for a train ferry across the stretch of water known as the Hamoaze, which comprised the estuary of the Tamar, Tavy and Lynher rivers. This bridge was opened in 1859 and consisted of two main spans of 455 feet, 100 feet above mean spring high tides. There is now a road bridge crossing the Tamar from Torpoint in Plymouth to Saltash, and as one drives across it there are excellent views of Brunel's rail bridge, and of the naval dockyards at nearby Devonport. Another notable suspension bridge that Barbara and I have admired in person is the Penang Bridge in Malaysia (see Plate 3(d)), of which more later.

But firstly, in relation to bridges more close to home, I must relate the story of my friend David Hughes' wedding. Since his wife Gloria was partly of German extraction, they were due to be married in the German Lutheran Church in Alie Street, Whitechapel. Since Gloria did not have any close male relatives to give the bride away, David asked me to perform this task. So I found myself in a large limousine with Gloria, driving to the Church. Unfortunately, the journey took less time than I had anticipated, and it being a great breach of etiquette for the bride to arrive before the groom, I instructed the driver to waste some time by driving over the adjacent Southwark Bridge, and then back over Tower Bridge (Plate 4(a)) to the Church. To my alarm, as we approached Tower Bridge, I saw that the keeper was about to raise the bridge to allow the passage of a vessel in the River Thames below. Since this, I knew, would mean a delay of up to half an hour, I leapt out of our car and persuaded the keeper to lower it for a few moments, so we could get across. The keeper's heart was soften by the sight of Gloria in her bridal finery, and complied with my request. To parody a well-known modern song, "we just made it to the church on time"!

There was a time when a previous version of London Bridge was being replaced and an American tycoon decided to buy the old version and transport it brick by brick to the USA. Unfortunately he thought he was buying the iconic and dramatic Tower Bridge, but, when the bridge he had purchased was put together again in Arizona, he was mortified to find that it was the much more mundane London Bridge that he had purchased! Tower Bridge (Plate 4(a)) was built between 1886 and 1894. The machinery for raising the halves (known technically as bascules) of the bridge to allow ships to pass is housed in the base of each of the two towers. The bridge was originally painted a greenish-blue colour, but was repainted red, white and blue in 1977 to celebrate the present Queen's silver jubilee. Tower Bridge is 800 feet long, and the towers are each 213 feet high. Each bascule can be raised to an angle of 86 degrees, and it takes 5 minutes to do this. The parts of the bridge on the bank sides of the towers, are suspension bridges, 270 feet long. The bascule-raising machinery is now electro-hydraulic, the water of the original hydraulic systems having been replaced by oil. There is the well-known story of a number 72 double-decker London bus being caught on the bridge when it started opening; the driver accelerated and just jumped the gap in time, and was awarded the George Medal for this life-saving feat!

The Golden Gate Bridge (Plate 4(b)) in San Francisco spans the Golden Gate, the opening of San Francisco Bay into the Pacific Ocean. This bridge was built to allow expansion of the rapidly growing San Francisco into Marin County on the opposite side of the bay, and was designed by one Joseph Strauss who had earlier designed (but not had built!) a 55 mile long bridge across the Behring Strait between Asia and Alaska. The Golden Gate Bridge is 8981 feet in length. The proximity of the Bridge to the San Andreas Fault makes it vulnerable to earthquakes, and it was constructed to withstand these seismic events. However, after it was opened in 1937, it was realized that its resistance to earthquakes was unlikely to be as good as previously thought, and improvements were instituted which were only completed in 2012.

The newest bridge in our personal collection is the Millennium Bridge in London (Plate 4(c)). This is, in fact, a purely pedestrian walkway, designed to celebrate the 2000 Millennium in the UK. Its north end is near St. Paul's Cathedral and the south end is in the Bankside district near the Tate Modern art gallery. Its length is 1,214 feet and width 13 feet. Unfortunately, when large numbers of people first used it, it began to wobble! This sort of resonance effect can eventually result in destruction of the bridge (and it is for this reason that marching soldiers are required to 'break step' when crossing vulnerable bridges). The characteristic feature of bridges at risk of such resonance is that they have lateral frequency modes of less than 1.3 hertz. So this bridge had to be closed for modifications only two days after it was opened, but it was safely re-opened in 2002. Other bridges with this problem are the Birmingham National Exhibition Centre, The Groves Suspension Bridge in Chester, and the Auckland Harbour Road Bridge, New Zealand.

To return to Brunel, he was appointed chief architect of the Great Western Railway in 1835, before the completion of the Rotherhithe tunnel. Brunel's hope was that passengers would be able to travel on one ticket from London's Paddington Station to New York, changing from the Great Western Railway to the '*Great Western*' steamship at the terminus in Neyland, South Wales. He made the decision to broaden the gauge of the line to 7ft. 2¼", because this would add superior running and greater stability at high speeds, and a more comfortable ride for the passengers; in addition, the wider gauge allowed for wider carriages and thus greater freight capacity. The Great Western Railway boasted striking viaducts, e.g. the one at Ivybridge, Devon (near where our daughter, son-in-law – David Legg, and family have just gone to live, following his appointment as a vicar of the Lee Mill parish). Paddington Station was also designed by Brunel and opened in 1854.

Simultaneously, Brunel was designing the first ship of the Great Western Steamship Company, named the 'Great Western', which was then the longest ship in the world (236 ft). It made its first crossing of the Atlantic in 1838, departing from Avonmouth, near Bristol, with 600,000 kg of coal, other cargo, and seven passengers aboard. However, it missed its opportunity to claim to be the first ship to cross the Atlantic under steam power alone, because a fire broke out aboard as it was returning from fitting out in London. Brunel had calculated that the amount a ship could carry increased as the cube of its dimensions, but the resistance due to water drag only increased by the square, meaning that a larger ship would consume less fuel per unit of cargo than a smaller one. The crossing took 15 days and 5 hours, and the 'Great Western' arrived at New York with a third of the fuel on board remaining unconsumed, demonstrating the reality of his theoretical calculations. As a result of this proving voyage, the Great Western Steamship Company set up a regular service between Bristol and New York, which operated from 1836 to 1846. The 'Great Western' was the first ship to hold the title of 'The Blue Riband of the Atlantic', with a crossing taking only just over thirteen days.

Brunel was convinced of the superiority of propeller-driven rather than paddle wheel-driven ships. He then turned his talents to travel to the Orient, designing the 'Great Eastern', which could carry over 4,000 passengers non-stop from London to Sydney, Australia. Though it took several decades for trans-oceanic steamship travel to become a commercial success, the 'Great Eastern' had, in fact, found another role as a telegraph cable layer, being involved in laying the first transatlantic cable which provided telecommunication between Europe and North America. There are a number of monuments to Brunel and his achievements, including statues at London Temple, and at Bristol, Saltash, Swindon, Paddington and Milford Haven stations. In addition, the building housing the Engineering Faculty at the University of Plymouth is named after him. His activities at Swindon made that a boom town, and he built houses, churches, recreational facilities ('The Mechanics Institute') and hospitals for the workers. Some claim that the Mechanics Institute was what gave Aneurin Bevan the impetus for the creation of the National Health Service. The remit of the Minister of Health at that time (the 1940's) also included Housing, and Bevan had to cope with a severe shortage of accommodation, because of the destruction of homes during the war by enemy air attacks, and due to the demands of the returning armed forces and the consequent increase in the population. Bevan famously said that in order to obtain the co-operation of the medical profession he "had to stuff their mouths with gold".

Returning to the subject of bridges, the other famous structure that we have visited is the Sydney Harbour Bridge. This connects the Sydney central business district with the North Shore of the Harbour and was opened in 1932. It is the world's widest single span bridge, being 160 feet wide, and has an upwardly convex steel arch spanning the two towers, from which steel rods support the road below. The total length of the bridge is about 2.4 miles. At the opening ceremony, the Premier of New South Wales was about to cut the tape, when a man on a horse galloped up and cut the ribbon himself; he was promptly arrested. There is a magnificent view of the Sydney Opera House from the Bridge; we have a special affinity with the Opera House, not because of the magnificence of the shows it puts on, but because in the foyer is a telephone. It is a tradition for all members of our family visiting that neck of the woods to call home from that telephone!

The next in our personal collection of bridges is the Penang Bridge (Plate 3(d)), which is a dual-carriageway toll bridge connecting George Town on the island of Penang off the west coast of peninsular Malaysia with Seberang Prai on the mainland. This bridge is 8.4 miles long and handles 65,000 vehicles daily. The bridge is six lanes wide and its construction was overseen by the Works Minister of the time, Dato Seri Samy Vellu, a friend of our Malaysian friends, the Ampikaipakans, and Barbara and I have met him on several occasions. The overall project was driven through by the Prime Minister of the day, Dr. Mahathir Mohamad, The bridge has two central pillars, from which the roadway is suspended by a series of steel cables, as in the Golden Gate Bridge. The Penang Bridge opened in 1985, and this was the first such decorative bridge of its kind, being designed by Chin Fung Kee.

The other way of crossing rivers with your car is by ferry. Barbara has a friend, Stella Butler, the widow of a High Court Judge, who has a cottage at the Cornish village of Fowey, which is situated on the Fowey River. Almost immediately opposite the cottage, on the opposite side of the river, is the hamlet of Bodinnick, and for many years a ferry has crossed the river at this point, designed to take both passengers and cars. In our early visits to Fowey, the ferry consisted of a motor boat with a large platform moored alongside it. It was possible with extreme care to drive your car on to this platform, and the whole contraption would set off (with ourselves in the motor boat) across the river, which had a very strong current; one felt that the whole thing was in danger of capsizing at any moment. Symbolically, during the crossing you were sold your ticket as you passed the moored local lifeboat. The whole *'Heath-Robinson' set-up has now been replaced by a more substantial ferry.

The next bridge I have visited whose features are of particular interest is the Charles Bridge in Prague. This crosses the Vltava River, and was built over a considerable period at the end of the 14th century. It connects Prague Castle with the city's Old Town and became part of an important trade route between Eastern and Western Europe. The most striking feature is the set of thirty baroque statues on the balustrades, placed there between 1683 and 1714. These include representations of St. Luthgard, the Holy Crucifix and Calvary, and St. John of Nepomuk. St. Luthgard (1182-1246) was a Flemish saint, born in Tongeren, in what is now Belgium. Because her potential dowry had been dissipated in her parents' failed business ventures, she could not get married and so chose the cloister as a socially acceptable alternative to the disgrace of being unmarried. She was therefore admitted to the Benedictine monastery of St. Catherine at the tender age of twelve. Here she was visited with a vision of Jesus who showed her his wounds (i.e. the 'stigmata') and at the age of twenty she became a fully-fledged Benedictine nun. She apparently experienced ecstasies, levitations, and dripped blood from her hair and scalp when in an entranced state. She asked Christ for a better grasp of Latin than she already had so that she might better comprehend the Word of God. She was, however, not satisfied, and returned to Christ asking if she could exchange the gifts she already had for something else. On being asked what she would like instead, she said "Lord, I would exchange them for your heart". Christ then reached into Luthgard, removed her heart, and replaced it with his own, hiding her original heart within his breast. To most people in the medical profession today, all this would probably add up to severe symptoms of schizophrenia!

The other saint I have mentioned in the last paragraph is John of Nepomuk (1345-1393). He is a national saint of the Czech Republic, who was drowned in the Vltava River at the order of Wenceslaus, King of the Romans and of Bohemia. This was because he refused to divulge the secrets of the confessional. This punishment was inflicted was because of a row about the appointment of a new abbot for the

Benedictine Abbey at Kladruby. This was the time of two Popes, one based at Rome and the other at Avignon. Wenceslaus backed the Avignon faction, whereas the Archbishop of Prague supported the Pope in Rome. Wenceslaus was afraid that his wife had taken a lover, whose identity she refused to disclose. King Wenceslaus therefore ordered John of Nepomuk to be drowned. Readers may wonder why Wenceslaus became known, as in the Christmas Carol, as 'Good King Wenceslaus'. This relates to a quite separate incident when the King braved harsh winter weather to give alms to a poor peasant on the Feast of Stephen.

One of the most famous bridges is Le Pont Saint Bénezet at Avignon, France. Its construction was inspired by a local shepherd boy (Bénezet) who was 'inspired by angels' to build it. He proved his divine inspiration by lifting a large block of stone! The bridge originally spanned the River Rhône between Avignon and Villeneuve-lès-Avignon and was built between 1177 and 1185, and is 900 metres in length. It fell into disrepair in the 17th century and lost about four of its arches; the remains of the ruin still stand. When Bénezet died he was interred in a small chapel standing on one of the bridge's surviving piers. The Rhône boatmen used the original chapel as a site for worship, but the presiding clerics refused to conduct services in it for fear it might collapse, so in the 18th century a new chapel was built on dry land at the foot of the bridge on the Avignon side. The bridge was originally of great strategic importance since it was the only Rhône road crossing between Lyon and the Mediterranean. The reason why this bridge is so well-known today is, no doubt, because of the song '*Sur le Pont d'Avignon*', composed in the 16th century by Pierre Certon, but with a different melody from the modern version, which dates from the mid-19th century when it featured in an operetta by Adolphe Adam entitled '*L'Auberge Pleine*'.

Finally, the sight of the ruined bridge at Avignon always reminds me of the ancient maritime punishment of 'walking the plank'. This was practised by pirates and mutineers, for the 'amusement' of the perpetrators. Captives were forced to walk off a wooden plank (this is where the analogy with the bridge at Avignon comes in – since both that bridge and the plank come to an abrupt end!)

Bridges also have featured in some well-known literary works, such as '*The Bridge over the River Kwai*', a novel written in 1952 by Pierre Boulle. It concerns British soldiers in a Japanese prisoner of war camp in western Burma, and is roughly based on fact.

The commandant, Colonel Saito, informs them that they are to work on the construction of a rail bridge over the River Kwai, but is reminded by the senior British Officer, Colonel Nicholson, that the Geneva Convention exempts officers from manual labour. Despite threats by Saito to shoot him, Nicholson refuses to back down, and is locked in a hole in the ground, without food or water. The non-commissioned men amongst the prisoners work at a deliberate snail's pace, and sabotage whatever they can. Saito has a deadline to meet, and, if he fails, will be forced to commit ritual suicide by personal disembowelment. He finds an excuse to issue a general amnesty for the prisoners. Meanwhile, Nicholson, a somewhat curious character, is shocked by the sloppy work done by his men, and orders some of his men to build a decent bridge, to maintain their morale, and so as to exemplify the ingenuity of the British Army. In the film, British Commandos, led by a Major Warden, are parachuted in, and at night when the tide is out, affix explosives to the bridge piers. A train carrying important Japanese dignitaries is due to cross the bridge the following day, so Major Warden plans to wait and destroy both the train and the bridge simultaneously, but he and his men are horrified to see that the water level in the river has dropped, exposing the detonating mechanism for the enemy to see. Nicholson doesn't want his masterpiece to be destroyed, so brings the mechanism to Saito's attention, and they hurry down to the riverbank to investigate. When the commandos reveal their mission to Nicholson, the latter at last realises the situation, and stumbles towards the detonator, but eventually falls on the plunger, destroying both the bridge and the train, and also himself. The whole episode was turned into this film. The incidents in it are largely fictional, though it correctly depicts the appalling conditions suffered by British prisoners at the hands of the Japanese. It was loosely based on the construction of the Burma-Siam railway by Commonwealth, Dutch and American prisoners of war. Saito was later tried by a war crimes tribunal, and was defended by Lieutenant Colonel Toosey, the real counterpart of Colonel Nicholson. Saito was, in reality, respected by his prisoners for being comparatively merciful and lenient towards them. The film music featured the '*Colonel Bogey*' march, which was later used for the song '*Hitler has only got one ball*'; however, the director, Sam Spiegel, thought the words to be too vulgar, so whistling was substituted for the actual song.

Another book about a bridge, this time more firmly based in reality, is '*A Bridge Too Far*', written by Cornelius Ryan in 1974. It is an account of the failure of '*Operation Market Garden*' during World War II, an Allied attempt to break through German lines and seize the Rhine bridges in the Arnhem area of the Netherlands. There was a rivalry between the US General George Patton and the British Field Marshal Bernard Montgomery, to be the first to get to Berlin. The overall Supreme Allied Commander, General Dwight D. Eisenhower, chose Montgomery's plan, which involved parachuting 35,000 men behind enemy lines in the Netherlands, the British having said that the terrain near Arnhem is ill-suited for the landing of transport aircraft bringing in essential supplies. They also pointed out that they would have to land in an area eight miles from the bridge and gliders would therefore have to be used to bring in Jeeps. The Arnhem Bridge was a crucial one, being the last means of escape for the German forces in the Netherlands. The operation was beset by technical problems, including failure of timely arrival of the gliders, and the radio sets turning out to be useless, so no that communications could be made between the various parts of the operation. The result was a failure to reinforce the British contingent. After days of house-to-house fighting in Arnhem between the paratroopers and the crack German SS infantry and Panzer divisions, many of the British paratroopers were either captured or forced to withdraw. The British Commander, Major-General Urquhart, managed to escape with fewer than 2,000 of his troops, the remainder being taken prisoner. The deputy commander of the First Allied Airborne Army, Lieutenant-General Sir Frederick Browning, on being asked by Urquhart for his opinion about the operation, replied "Well, as you know, I always felt we tried to go a bridge too far", thus providing the subsequent film with its title.

Finally, in this account of bridges in literature, we come to 'The Bridge of San Luis Rey', American author Thornton Wilder's Pulitzer Prize-winning novel about several people who die in the collapse of an Inca rope bridge in Peru, and the events which led up to them being on the bridge at that fatal moment. There are overtones of fate and destiny throughout the book. The collapse was witnessed by one Brother Juniper, a Franciscan monk, who was himself on his way to cross it. The first of the victims whose story is told is the Marquesa de Montemayor, whose daughter, Clara, subsequently, much to her mother's distress, had vanished to Spain. So the Marquesa had taken as a companion a certain Pepita, a girl raised at a local convent. When the Marquesa is out visiting a local shrine, Pepita writes to the Abbess of the convent, complaining about her loneliness and general misery. The Marquesa sees the letter on her return, and, not realising that the Marquesa has read it, Pepita says she has burnt it, because it was not brave to write it. The Marquesa has also been given new insights into the way her own life has lacked bravery. Two days later, the Marquesa and Pepita are on the bridge when it collapses.

The second victim is Esteban, who, with his twin brother Manuel, was left at a convent when they were infants. They are very close to each other, but the relationship becomes strained when Manuel falls in love with one Camila Perichole. Camila's lover happens to be the Viceroy of Peru. When Esteban eventually discovers this, he gets his brother Manuel to write to a bullfighter with whom she is also having an affair. Esteban encourages Manuel to follow her, but the latter swears that he will never see her again. Shortly afterwards, Manuel dies from an infected cut. When the Abbess of the convent comes to lay out the body, she asks Esteban his name, and he replies "Manuel". Gossip about his strange behaviour is widespread, and the Abbess sends for a Captain Alvarado, who hires Esteban to sail with him. Esteban wants his pay in advance to buy a present for the Abbess, and journeys towards Lima with the object of buying the present there. At the ravine, the captain goes down to a ferry boat that is crossing the river, but Esteban decides to go via the bridge and is on it when it collapses. Brother Juniper works for six years on his book about the collapse of the bridge, trying various mathematical formulae to measure spirituality, but without success. A special council declares his book to be heresy and it and Brother Juniper are burnt at the stake.

Thornton Wilder, the author of the book, said that in it he was posing the question of whether there is a direction and meaning in lives beyond the individual's own free will. The bridge itself, in both Wilder's novel and Prosper Merimee's subsequent play based on it, is based on the great Inca road suspension bridge across the Apurimac River, erected in 1350 and still in active use in 1864, in a somewhat dilapidated state in 1890.

The third victim is Uncle Pio, who is Camila Pericholes' singing master, coiffeur, masseur, errand boy and banker. He has travelled widely, on business related to the theatre or politics, and conducting interrogations for the Inquisition. He discovers a twelve year old café singer, who develops into Camila Perichole, who later has three children fathered by the Viceroy. When a smallpox epidemic sweeps through Peru, Camila catches the disease and becomes disfigured by it. Uncle Pio sees her trying unsuccessfully to cover her pock-marked face. He begs her to be allowed to teach a boy (Jaime), who is travelling with him, as he taught her. They leave next morning, and thus Uncle Pio and the boy Jaime become the fourth and fifth people on the bridge to Lima when it collapses.

Returning to the whole subject of going down, rather than up, it is first necessary to consider the possible limits (I have described later a journey to the centre of the earth, according to *Jules Verne!). The greatest known depth (about 36,000 feet) on Earth is the Mariana Trench, in the Pacific Ocean, near the Northern Mariana Islands. The British vessel HMS Challenger II surveyed the Trench in 1951. The Earth is the only one of the planets known to have oceans, which cover approximately 72% of the Earth's surface. The total volume of oceans on earth is about 1.3 billion cubic metres, and the average depth is 12,080 feet. Oceans are the habitat of 230,000 *known* species, though there are vast numbers of undescribed creatures. The only other body in the Solar System known to have surface water is Titan, a satellite of Saturn, which is mainly composed of water, ice and rocky material. It was discovered in 1655 by the Dutch scientist, mathematician and astronomer, Christiaan Huyghens (1629-1695).

The word 'bridge' has also featured in many other ways in the English language. When you want to buy a house, you are usually required to pay a substantial proportion of the cost at the time of signing the contract of sale – a sum which you very likely cannot afford. Your bank is usually willing to give you a 'bridging loan' to cover the period between the signing of the contract and the completion of the deal. The only downside is that bankers usually charge pretty high interest rates for this sort of loan, so it is prudent to 'complete' as soon as possible. Many children have rather crooked teeth, which may be straightened out by constructing bridges between them, or by wearing a dental plate for a few weeks.

Another use of the word 'bridge' is in the card game 'contract bridge', a game to which my parents and their friends in Plymouth, their home town, were deeply addicted, much to my own personal boredom. Some players have risen to great fame because of their expertise at the game, notably the actor Omar Sharif. He was born in 1932 into a Lebanese Christian family, and graduated from Cairo University with degrees in mathematics and physics, and later starred in such notable films as David McLean's '*Lawrence of Arabia*' and '*Dr. Zhivago*'. He also appeared in the title role in '*Genghis Khan*', and as a German general in '*The Night of the Generals*'. He ran a team of outstanding bridge players – the 'Omar Sharif Bridge Circus' which toured the world, competing against other teams with great success. When King Farouk of Egypt was deposed by Gamel Abdel Nasser (1918-1970), restrictions were imposed on foreign travel by the new regime, so Sharif decided to remain in Europe, where he happened to be at the time of the coup. Nasser's relationship with the West suffered because of his neutralist policies in the Cold War, which had led to the withdrawal of promised Western funds for the construction of the Aswan High Dam. Nasser retaliated in 1956 by nationalising the Suez Canal Company. The Western powers could not tolerate this, since they relied on passage through the canal for much of their trade with the Far East. So The United Kingdom, France and Israel invaded the Sinai region and took control of the Canal, leading to a political fiasco, largely because the operation did not have the active support of the USA. Nasser instituted a number of reforms in Egypt, including establishing universal health care, expanding women's rights, family planning programs and housing. He became head of the Non-Aligned Movement, but in circumstances where opposition was banned. Tensions between Israel and the Arab states led to the Six Day War of 1967, in which the Arabs were defeated and Israel occupied much Arab territory. Nasser resigned, but was re-instated by popular demand. Nasser died of a heart attack during a conference in which he was attempting to broker an agreement between the Palestine Liberation Organisation, led by Yasser Arafat, and Jordan. Meanwhile, Omar Sharif was awarded the Sergei Eisenstein Medal by the United Nations Educational, Scientific and Cultural organisation in recognition of his contributions to cinema and to cultural diversity.

To return to the 'depths', deep sea exploration used vehicles named 'bathyscaphes'. The first was built in the late 1940's by Auguste Piccard, and in 1960 a third version, the '*Trieste*', carrying Auguste's son Jacques and a Lieutenant Don Welsh reached the deepest part of the Mariana Trench. Bathyscaphes have floats rather than being suspended by a cable and they clearly have to withstand massive pressures. To descend, the vehicles air tanks are filled with sea water. To ascend, and since it is impossible to expel the sea water at the depths reached, ballast in the form of iron shot is released. The bathyscaphe, not being suspended from its mother vessel on the surface, has motors which allow it to make sideways journeys, power being provided by electric motors, fuelled by batteries. The vehicles were equipped with powerful lights, and these were able to pick up live creatures, such as diatoms, and 'some type of flatfish, resembling a sole'.

Submarines
Ancient art depicts men using schnorkel-like devices to swim under water, but clearly the depth attainable was limited. Legends from the 12th century AD (deriving from

the pen of Aristotle) suggested that Alexander the Great himself was lowered in 413BC in a primitive submersible to inspect the enemies defences during the siege of Syracuse. This vehicle was a form of diving bell, according to a 16th century Islamic painting. The strategic advantages of submarines were enumerated by Bishop John Wilkins of Chester in 1648 in a volume named 'Mathematical Magick'. It is surprising that this cleric, who should have been engaged in matters spiritual rather than military, should have been thinking about submarines! However, clerics of this era were often engaged in matters other than religious (e.g. Cardinal Wolsey (1473-1530), who apart from his religious duties, dealt with almost all matters of state, including foreign policy). Bishop Wilkins set out the advantages of submarines, such as:- privacy, safety from tides and tempests, pirates, and the secret blowing up of enemy vessels in time of war. Between 1690 and 1692, the French physicist Denis Papin designed and built two submarines. However, the Turkish Radio Television Corporation has claimed that the world's first submarine was invented by Turks in 1719. In the 17th century, Ukrainian Cossacks were using a submersible named a '*Chaika*' (gull) for reconnaissance and infiltration. The first real military submarine, the '*Turtle*', was designed by the American, David Bushnell, and crewed by a single man. During The American War of Independence, the '*Turtle*' tried, but failed, to sink a British warship, HMS *Eagle*, in New York harbour. During the American Civil War (1860-1865) the Confederate States fielded several human-powered submarines, for attacking Union ships that were blockading Confederate seaports. They had a limited air supply – i.e. only that which was contained in the vessel before it submerged. The first time a submarine successfully sank another ship was in 1864, when the submarine *Hunley* sank USS *Housatonic* off Charleston Harbour. All these submarines relied on human power for lateral propulsion, but from 1863 onwards engines were installed for that purpose.

The first time submarines had a significant impact on the outcome of a war was in World War I, when German U-boats attacked Allied commercial vessels. Aircraft carriers were equipped with waterproof hangars and steam catapults for launching seaplanes. The British built HMS *M2*, the French a vessel named '*Surcouf*', and the Russians started the war with 58 submarines in service. The Treaty of Versailles banned Germany from having submarines, but their construction started in secret in the 1930's. Eventually, this became known, and the Anglo-German treaty of 1936 allowed Germany to achieve parity with the UK in the number of its submarines.

The real age of the submarine was World War II, when Germany switched its major warship building to the U-boat (*Unterseeboot*) and used them to devastating effect in the Battle of the Atlantic. Britain was highly dependent for supplies of food and for parts for industry on imports. The 'wolf-packs' of German submarines communicated with each other using messages encrypted by the Enigma cypher machine, whose code was famously worked out by Alan Turing. The U-boats put to sea independently, and searched for convoys, which they then shadowed, and radioed to the German High Command to allow other U-boats in the region to converge and attack the convoy simultaneously. The Japanese had a large fleet of submarines, some of which had manned torpedoes. In 1942 they sank two Allied aircraft carriers, one cruiser and several destroyers and other warships. They were unable to carry on so successfully once the allied fleets became better organised. Britain had 70 submarines, of various classes, which operated in the North Sea and off the coast of Norway. They also attacked Axis supplies to North Africa from their base in Malta, which was severely bombed; the island as a whole was awarded the George Cross for its efforts and the bravery of its inhabitants. British submarines also operated in the Pacific against the Japanese. The USA had 314 submarines, and lost 52 of them to depth charges released by the destroyer escorts of the Japanese convoys. Early on in the war, the Japanese triggered their depth charges at too shallow depths to be effective against Allied submarines. Post-war, a number of other submarine developments took place, including modifications to allow the use of cruise missiles, some fitted with nuclear warheads. Some submarines are now nuclear-powered, allowing them to remain at sea, submerged, for very long periods, and this allowed the crossing of the North Pole underwater. The latest use of a nuclear-powered submarine was in the Falklands War of 1982, when the Argentinian cruiser *General Belgrano* was sunk by HMS *Conquerer.*

During World War II, the Germans stationed the battleship *Tirpitz* in Norway, in order to attack convoys bound for the Soviet Union, and to prevent an invasion of Norway, which was then occupied by Nazi forces. *Tirpitz* was formidably armed with both guns and torpedo tubes, and its steel plates were up to 120 mm thick. After docking at Wilhelmshaven, a move designed to conceal her actual destination, she was moved to Trondheim Fjord in Norway. She was heavily camouflaged with net and foliage, in order to deceive prying Allied reconnaissance planes. Her freedom of action was, however, severely restricted by shortages of fuel, and the withdrawal of the German destroyer force to support the movements of the heavy warships

Scharnhorst, Gneisenau and *Prinz Eugen* up the English Channel. This assemblage of German naval might was intended to intercept convoys sailing to and from northern Russian ports. To provide an exercise in mutual cooperation, these forces were ordered by Admiral Karl Doenitz to attack Spitzbergen, which hosted a British weather station and refuelling base. This base was duly attacked by *Tirpitz's* artillery, which destroyed its facilities.

This was, in fact the only time the *Tirpitz* fired her guns in anger. At the end of September, 1943, British mini-submarines slipped under the torpedo nets and dropped two powerful mines under the ship, which extensively damaged it, but not beyond repair. It was then attacked by Barracuda dive-bombers, but again the ship was patched up. Numerous other attacks again failed to put the *Tirpitz* out of action permanently. Eventually, however, it was destroyed in Operation Catechism on 12th November, 1944, in which a force of Lancaster bombers blew a large hole in the ship's side, which then gradually turned over and capsized.

Another successful British raid took place in December, 1942, on the submarine pens at the mouth of the River Gironde, in Bordeaux harbour. This was the subject of a film made in 1955 entitled '*The Cockleshell Heroes*', starring Trevor Howard and José Ferrer. This was a commando operation designed to plant limpet mines underwater on the sides of many ships in the harbour. The raid was successful, and many ships destroyed, but only two of the marines managed to escape, the rest being captured, and shot by the Germans when they refused to divulge the nature of their mission.

Lloyd George

David Lloyd George (1863-1945) was a fiery Liberal politician and statesman, best known for guiding Britain through the First World War, and as leader of a coalition government between 1916 and 1922. He is the last Liberal to have served as Prime Minister, and is regarded as the founder of Britain's welfare state. He was first elected to Parliament in 1890 as Member of Parliament for Carnarvon Boroughs. He campaigned to disestablish the Church of England, for temperance reform and for Welsh Home Rule. He attacked Joseph Chamberlain, accusing him of profiteering through the Chamberlain family company, Kynoch Ltd, of which Chamberlain's brother was the Chairman, and which had won contracts issued by the War Office, even though its prices were higher than those of competitors. In 1906 he entered the liberal Government of Sir Henry Campbell-Bannerman as President of the Board of Trade. He then became Chancellor of the Exchequer, with a commitment to reduce military expenditure. In 1909 he introduced a Budget imposing increased taxes on luxuries such as liquor, tobacco and land, to raise money for new welfare programmes. He was an opponent of war until the Agadir crisis of 1911, but then made a speech attacking German aggression, and because Belgium was 'a small nation, just like Wales'. In 1915 he became Minister of Munitions, and then Secretary of State for War when Lord Kitchener was drowned on his way to Russia. He insisted that there must be a 'fight to the finish', thus rejecting American president Woodrow Wilson's offer to mediate. The Prime Minister of the time was Asquith, who was generally regarded as a poor planner and organiser, and after he had refused to agree to Lloyd George's demand that he be allowed to chair a small committee to manage the War, Asquith was forced out and succeeded by Lloyd George, with much popular support. A cartoon in *Punch* showed him as 'The New Conductor' conducting the orchestra in the '*Opening of the 1917 Overture*', an allusion to the Russian revolution. One of Lloyd George's principal objectives was the destruction of Turkey, which had entered the War on the side of the opposing powers, and then to capture Jerusalem, to impress British public opinion. He brought nearly 90% of British merchant shipping under state control, and raised the matter of organised convoys on the transatlantic route. This was a solution proposed by Admiral Lord Jellicoe in response to the UK's increasing losses. Lloyd George also played a significant role in Arthur Balfour's declaration in favour of the establishment in Palestine of a national home for the Jewish people. Victory in the First World War came on 11th November, 1918, a date still commemorated in the UK by the sale of poppies, with the proceeds going to the support of wounded war veterans. At the Versailles peace conference of 1919. Lloyd George opposed the wishes of French premier Georges Clemenceau, US President Woodrow Wilson and Italian Prime Minister Vittorio Orlando, who all proposed to exact punitive reparations. Lloyd George did not want to destroy the German economy and political system by demanding massive reparations, and, in view of the later rise of Adolf Hitler, who describes his anguish at the Versailles accords in '*Mein Kampf*', ('*My struggle*'), the first page of which rather unpleasant book states "Austria must be returned to the great German Motherland", Lloyd George was probably right to take this attitude.

Lloyd George then had to deal with the perennial Irish problem. He called an Anglo-Irish convention in 1917 to try to deal with the demand for Home Rule for Ireland,

Plate 1(a). Jacob's Ladder, by William Blake

Plate 1(b). Statue, Edgware Road Station, London

Plate 1(c). Marsyas, a copy from an unknown source.

Plate 1(d). Ladder carving, with ascending figures. (From the west façade of Bath Abbey)

Plate 2(a). Deposition, by Albrecht Dürer

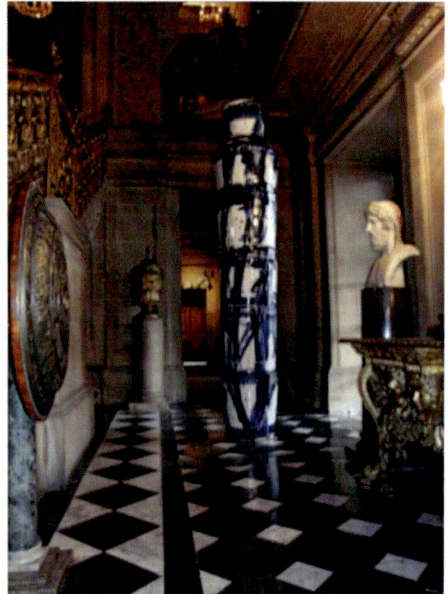

Plate 2(b) A glazed porcelain ceramic by Felicity Aylieff (born 1954) at Chatsworth House, Derbyshire

Plate 2(c). Een boerin die vlas kneust (naar Millet), (Vincent van Gogh, 1889)

Plate 2(d). The Nativity, Piero Della Francesco (1460-1475)

Plate 3(a). The Promenade (1917), Marc Chagall

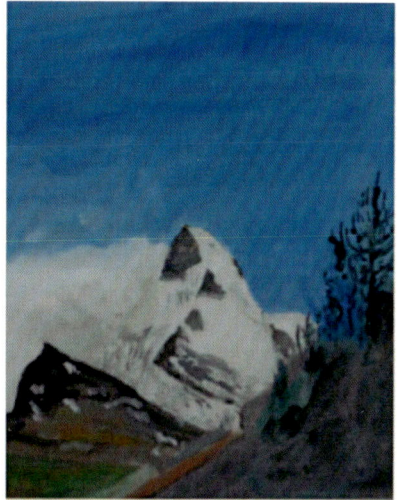

Plate 3(b). Painting of the Matterhorn (RDC)

Plate 3(c). The Clapper Bridge at Postbridge, Devon, UK

Plate 3(d). Penang Bridge, Malaysia

Plate 4(a). Tower Bridge, London, UK

Plate 4(b). The Golden Gate Bridge, San Francisco, USA

Plate 4(c). The Millennium Bridge, London, UK

but his attempt to extend conscription to Ireland in April, 1918, was disastrous, and the moderate Irish Home Rule Party was replaced by the much more militant Sinn Féin, which immediately declared Ireland to be a Republic. Lloyd George introduced the Government of Ireland act in 1920; this led to the Anglo-Irish Treaty in 1921, in which Northern Ireland was spilt from the rest of Ireland.

Lloyd George was said to be consistently pro-German after 1923. In 1936 he went to Germany to talk to Adolf Hitler, and returned saying that "the Germans have definitely made up their minds never to quarrel with us again" – a misjudgement of enormous proportions, reminiscent of that made by Neville Chamberlain (*vide infra*). A pessimistic speech he made in 1941 caused Churchill to compare him with Philippe Pétain.

Lloyd George also had a considerable reputation as a womaniser, giving rise to the famous ditty:-

'Lloyd George knew my father
My father knew Lloyd George
Lloyd George was my father!'
etc., etc.

Neville Chamberlain (1869-1940)

Chamberlain is best known for his policy of trying to appease Hitler, and, in particular, for the signing of the Munich Agreement in 1938, which conceded the Sudetenland region of Czechoslovakia to Germany. He became Minister of Health in the Government of Stanley Baldwin, and then Chancellor of the Exchequer in 1931. Chamberlain had had to deal with the 1936 Edward VIII abdication crisis. He agreed with Baldwin that if the King married the American divorcee, Wallis Simpson, he should abdicate, which he eventually did on 10th December. Both Chamberlain and Baldwin stressed that the King should make his decision before Christmas, since the uncertainty was harming the Yuletide trade! Chamberlain told the Cabinet:-

"It is now perfectly evident that force is the only argument Germany understands, and... heaven knows, I do not want to get back to alliances, but if Germany continues to behave as she has done lately, she may drive us to it"

His Premiership was, however, dominated by the question of what to do about Hitler. In February, 1938, Hitler began to pressure the Austrian Government to accept '*Anschluss*', i.e. union between Austria and Germany, which eventually invaded Poland on 3rd September, 1939.

Since the Sudetenland (the area on the northwest of Czechslovakia) contained the largest ethnically and linguistically German population (about 3 million), Hitler began to call for its union with Germany. Meanwhile, Britain and Italy signed an agreement in 1938, in which, in exchange for *de jure* recognition of Italy's Ethiopian conquest, Italy agreed to withdraw some of its volunteers from the pro-Franco (Nationalist) side of the Spanish Civil War. In May 1938, two Sudeten German farmers attempting to cross into Czechoslovakia without stopping for border controls were shot by Czech border guards, in response to which Germany was said to be moving troops to the border. Lord Halifax, the British Foreign Secretary, and Chamberlain were generally applauded for their handling of the crisis, and received almost unanimous support from the British press. On 30th August, Chamberlain met with his Cabinet and secured its backing for a policy of pressurising Czechoslovakia into making concessions because Britain was in no position to back up any threat to go to war over the issue. On 13th September, Chamberlain was informed by his secret services that Germany would invade Czechslovakia on 25th September. He then sent a message to Hitler that he was willing to come to Germany for further negotiations. On 15th September, Chamberlain flew to Munich and journeyed on to Berchtesgaden, Hitler's retreat. He received assurances that Hitler would have no further designs on the rest of Czechoslovakia other than the annexation of the Sudetenland, or on other areas of Eastern Europe which had German minorities. Chamberlain then flew back to Germany and met Hitler again in Bad Godesberg on 22nd September, but this time Hitler repudiated the proposals of the previous meeting, saying "that won't do any more". On 29th September, Chamberlain flew again to Germany for his third and final visit to Germany as Prime Minister, with the French Prime Minister, Daladier. On his return to the UK, Chamberlain took a piece of paper from his pocket, which was entitled 'Anglo-German Agreement'. The German Foreign Minister, von Ribbentrop, had remonstrated with Hitler for signing it, to which the latter replied "Oh, don't take it so seriously". When this news reached London before Chamberlain arrived back home, there was much dismay amongst Winston Churchill and his colleagues. Chamberlain appeared on the balcony of Buckingham Palace, and was asked to declare 'peace in our time', which he refused to do. But what he did say in a statement to the crowd was:-

"My good friends, this is the second time there has come back from Germany peace with honour. I believe it is peace for our time. We thank you from the bottom of our hearts. Now. I recommend you to go home, and sleep quietly in your beds".

Though most newspapers supported Chamberlain, Churchill was deeply opposed to what had happened, and told the House of Commons:-

"England has been offered a choice between war and shame. She has chosen shame, and will get war".

Nevertheless, Chamberlain did take some action, he signed a number of defence pacts with European countries, hoping this would deter Hitler from war, and informed the House of Commons of British and French guarantees that they would assist Poland if attacked. He doubled the size of the Territorial Army, created a Ministry of Supply to oversee the provision of equipment to the armed forces, and instituted peacetime conscription.

All this was, however, ineffectual. Germany invaded Poland on 1st September, 1939, in the early hours of the morning. At 11.15 a.m., Chamberlain addressed the nation by radio, telling them that it was now at war with Germany, saying that no word given by Germany's ruler could be trusted and the situation had become intolerable. (Both Barbara and I can remember receiving this news with shock – I was on holiday in Bournemouth with my parents at the time). Chamberlain set up a War cabinet, and invited the Labour and Liberal Parties to join his government, but they declined. He restored Churchill to the Cabinet as First Lord of the Admiralty, and also gave Anthony Eden a post. Churchill, however, could not keep quiet and sent a series of notes to Chamberlain advising him on how the war should be pursued. Nevertheless, Chamberlain, seeing the writing on the wall for himself, resigned and recommended that Lord Halifax should succeed him. Halifax was, however, reluctant, saying that he could not govern effectively from the House of Lords and Attlee said that Labour would not serve under Chamberlain, so the latter advised the King to send for Churchill. Chamberlain died of bowel cancer in 1940, and was eulogised by Churchill in the House of Commons:-

By 10th May 1940, it had become clear that the country had no confidence in Chamberlain, who promptly resigned. A meeting between Chamberlain, Halifax, Churchill and David Margesson, the Government Chief Whip, led to the recommendation of Churchill as Prime Minister, which King George VI, as a constitutional monarch, accepted.

Chamberlain, who died of bowel cancer in 1940, was eulogised by Churchill in the House of Commons:-

"Whatever else history may or may not say about these terrible tremendous years, we can be sure that Neville Chamberlain acted with perfect sincerity according to his lights, and strove with the utmost of his capacity and authority… to save the world from the awful devastating struggle in which we are now engaged".

Though some, including Halifax, favoured a negotiated peace with Germany, Churchill refused, and proceeded, in some of his finest speeches, to prepare the British for a long war. One of his first appointments was to put his industrialist friend, Lord Beaverbrook, in charge of aircraft production. He then told the House of Commons that he had nothing to offer except "blood, toil, sweat and tears", and he later declared that:-

"We shall fight in France, we shall fight on the seas and oceans, we shall fight with growing confidence and growing strength in the air, we shall defend our island, whatever the cost may be, we shall fight on the beaches, we shall fight on the landing grounds, we shall fight in the fields and in the streets, we shall fight in the hills; we shall never surrender", and "we shall brace ourselves to our duties, and so bear ourselves, that if the British Empire and its Commonwealth last for a thousand years, men will still say "this was their finest hour".

After the second Battle of El Alamein, Churchill declared:-

"This is not the end. It is not even the beginning of the end. But it is, perhaps, the end of the beginning".

Churchill capitalised on his good relations with the United States, and exchanged about 1,700 letters and messages with the US President, Franklin D. Roosevelt, and met him eleven times between 1939 and 1945. Roosevelt set about supplying Britain with oil, munitions and food via the North Atlantic shipping routes. Roosevelt persuaded his Congress that repayment for this service would, in fact, be defending the USA, and so 'Lend-Lease' was created. Churchill formed the Special Operations Executive, which established covert, subversive and partisan activities in enemy territories. When Hitler invaded the Soviet Union, Churchill, strongly anti-Communist, stated "If Hitler invaded Hell, I would at least make a favourable reference to the devil in the House of Commons". During October, 1944, he and the Foreign Secretary, Anthony Eden, visited Moscow to plan the post-war Europe, and who would have have which territory within its sphere of influence. One of the conferences was at Yalta, on the Black Sea, at which some of the principles established were:-

Agreement that Nazi German surrender should be unconditional.

Berlin should be divided up into four occupation zones – American, British, French and Russian.

German reparations were partly to be in the form of forced labour, to repair the damage done by Germany.

Stalin wanted all sixteen Soviet Socialist Republics to get separate United Nations membership, but this request was denied, presumably because it would have grossly unbalanced the voting power at the United Nations.

Stalin agreed to enter the war against Japan within 90 days after the defeat of Germany.

Nazi war criminals were to be hunted down and brought to justice.

Polish troops – over 200,000 in strength, agreed to serve under the overall command of the British Army, and were instrumental in the defeat of the Germans in North Africa and Italy. However, at Yalta, Roosevelt and Churchill largely gave in to Stalin's demands that he should annex territory which in the Nazi-Soviet Pact had been designated to go under Soviet control. In consequence, many thousands of Polish troops lost their homes to the Soviet Union, and, in reaction, about thirty officers and men from the Polish II Corps committed suicide. Many Polish troops refused to return to their homeland, because of Soviet repression of Polish citizens, and provision was eventually made for them to stay in Britain in the Polish Resettlement Act of 1947.

To return to Roosevelt, when he took office in 1933 a majority of the nine judges of the Supreme Court had been appointed by Republican Party Presidents, and many were especially conservative. Roosevelt asked Congress to approve an Act which made retirement from the Supreme Court obligatory when the age of 70 was reached. But he did not reckon with many congressmen who feared that he might next start to retire *them* at the age of 70 and the proposal failed. Roosevelt died whilst in office in 1945 and was succeeded, according to the provisions of the US Constitution, by the Vice-President, Harry S. Truman. President Truman's most important decision was to use the atomic bomb against Japan to finish the war off quickly; the first was dropped on Hiroshima on 6th August, 1945, and the second three days later on Nagasaki. Hiroshima was a supply and logistics base for the Japanese, had large stores of military equipment and was a key port. Nagasaki was a large sea port in southern Japan, and had wide-ranging industrial activity related to the prosecution of the War. The first bomb killed an estimated 90,000-160,000 people and the second 60,000-80,000 from the acute effects alone. Later deaths were due to burns, *radiation sickness, and to the development of certain types of cancer. One of my memories from early childhood was a newspaper photograph taken in one of these two cities which showed a faint darkening on a wall, the shape of a human body. What had happened is that a man standing in front of the wall had simply evaporated, with the darkening being due to the slight lowering of temperature due to shielding of the wall by the man's body. There were extensive debates about the morality of these attacks, but the general view was that they shortened the War, thus preventing even greater loss of life. On 15th August, six days after the bombing of Nagasaki, Japan surrendered.

President Truman strongly supported the creation of the United Nations, and included Roosevelt's widow, Eleanor, in the delegation to the United Nations first General Assembly. He also supported the Marshall Plan which aimed to help Europe rebuild its economy. He responded to the Soviet Union's attempt to block access to the three Western-administered zones of Berlin, and by instigating the Berlin air lift, in which massive supplies of food and fuel were flown daily into the beleaguered city, escorted by military aircraft. He also was a supporter of the plan for a Jewish homeland in Palestine, writing that "Hitler had been murdering Jews right, left and centre... The Jews need some place to go. It is my attitude that the American government couldn't stand idly by while the victims of Hitler's madness are not allowed to build new lives". In June, 1950, Kim Il-sung's North Korean forces invaded South Korea and Truman urged the United Nations to intervene, which it did, and US forces under United Nations auspices, and led by General Douglas MacArthur, landed at Inchon and marched north towards the Yalu River boundary with China; this produced a massive response from China, and by early 1951, stalemate had been reached at the 38th parallel of latitude, i.e. where the war had begun. MacArthur leaked his plan of attacking Chinese supply bases north of the Yalu to the Press. Since Truman was worried that further escalation of the war might lead to open conflict with the Soviet Union, he dismissed MacArthur from his commands. One of MacArthur's melodramatic statements on his dismissal was:-

"Old soldiers never die, they merely fade away!"

In 1950, an attempt to assassinate Truman was made by Puerto Rican nationalists, who were trying to achieve independence for their country. They managed to mortally wound a White House policeman, but Truman allowed a plebiscite to take place in Puerto Rico, in which 82% of the electorate voted for independence. Another of his achievements was to achieve equal status with whites for African-American soldiers on their return from World War II, and in due course the various branches of the military became racially integrated. Truman died in 1972, from lung cancer.

Joseph Stalin (1878-1953)
Stalin is a good candidate for award of the palm as the nastiest person in history. He was the son of Besarion Jughashvili, a cobbler, and Ketevan Geladze. In childhood he was plagued by a number of illnesses, and was left permanently scarred after an attack of smallpox. He and his mother were subjected to physical abuse by his father, an alcoholic. He was banished from the town of Gori, in present-day Georgia, after assaulting a police chief, and subsequently moved on his own to Tiflis. He got a scholarship to a Georgian Spiritualist Seminary in Tbilisi, but was expelled for missing his final exams. At about this time he discovered the writings of Lenin and decided to become a Marxist revolutionary, joining the Bolsheviks in 1903. He

organised numerous strikes, and wrote an article about the first anniversary of the march in 1905 to the Winter Palace. He and his colleagues also organised paramilitaries, spread propaganda, and raised money through bank robberies, kidnapping ransoms and extortion. He attracted the attention of the secret police (the Okhrana) and he was sent to Siberia on seven occasions. Although conscripted by the Russian army to fight in World War I, he was deemed unfit for service because of a deformed arm. At the Communist Party conference in 1917, he was elected to the Bolshevik Central Committee, which led the insurrection against the Kerensky government in the October Revolution. Lenin formed a five-man Politburo, which included Stalin and Trotsky. However, Stalin challenged many of Trotsky's decisions, ordered the killing of many counter-revolutionaries and former Tsarist officers, burned villages to intimidate the peasantry, and in May, 1919, in order to stem desertions on the Western Front, had deserters publicly executed as traitors. When Lenin died of a heart attack in 1924, a power struggle began between the seven members of the Politburo, who included both Stalin and Trotsky. Because of a serious shortage of grain supplies in 1927, Stalin pushed for the collectivisation of farms. Bukharin and Rykov were removed by Stalin and his allies, the latter being replaced by Vyacheslav Molotov, who later became Minister of Foreign Affairs (1939-1949); however, he too fell out of favour with Stalin and was replaced by Andrei Vyshinsky.

A strong vein of anti-Semitism ran through Stalin's thinking, and he ordered Molotov to purge the government of Jews, believing that every Jew was a potential spy. In his paranoia on this subject, he ordered the development of bombers capable of reaching the USA, convinced that Harry Truman was Jewish. He said "we will show this Jewish shopkeeper how to attack us!" Since many of his doctors were Jewish, he refused to be treated by doctors, and would only be treated by veterinary surgeons, which seems appropriate for the great 'beast' that he was! In 1952-1953, Stalin accused a group of Moscow doctors, who were mainly Jewish, of plotting to assassinate Soviet leaders – the so-called 'Doctor's Plot'.

In 1968 Czechoslovakia, under its reformist leader, Alexander Dubček, who endeavoured to increase democracy, decentralise the economy, and loosen restrictions on the media, on travel and speech ('The Prague Spring'), oversaw the splitting of Czechoslovakia, which was heavily under the control of the USSR, into the Czech and Slovak Republics. The Prague Spring inspired the works of many authors, such as Václav Havel and Milan Kundera. The Soviet Union could not, however, tolerate this liberalisation, and in August, 1968, armies from the Soviet Union, Bulgaria, Poland and Hungary invaded what had been Czechoslovakia. The Soviet Union justified this response by invoking the 'Brezhnev Doctrine', which stated that the Soviet Union had the right to intervene whenever a country in the Eastern Bloc appeared to be moving towards capitalism. In Czechoslovakia, protests took place in several ways. Jan Palach, a student, set fire to himself in Wenceslaus Square, people gave wrong directions to invading soldiers, and others identified and followed cars belonging to the secret police. Dubček was arrested and taken to Moscow for negotiations, where, surprisingly, it was agreed that a programme of moderate reform would remain in place, and that Dubček would remain in office. In April, 1969, Dubček was replaced by Gustav Husák, and was given a job as a forestry official. Husák reversed Dubček's reforms, the only significant change remaining being the split into the Czech and Slovak Socialist Republics. (During 1987, Mikhail Gorbachev acknowledged that his policies of *'glasnost'* and *'perestroika'* owed much to Dubček's 'socialism with a human face'.) Until now, there were many left-wing people in the West who had admired Soviet views and policies, but the events of the Prague Spring caused disillusionment amongst them.

Anthony Eden (1897-1977)
Eden was a British Conservative politician who was best known for his role as Foreign Secretary for various periods between 1935 and 1955. He supported a policy of non-interference in the Spanish Civil War, and also Neville Chamberlain in his efforts to preserve peace by concessions to Hitler. The Italian-Ethiopian War was in the offing and Eden vainly tried to persuade Mussolini to submit the dispute to the League of Nations. Mussolini scoffed at Eden, describing him as "the best dressed fool in Europe". Eden resigned in 1938, over Chamberlain's policy of appeasement. Eden did, however, become Prime Minister in 1955 and but then resigned over the Suez crisis in 1957 because of criticism both at home and in America. This crisis was caused by the nationalisation of the Suez Canal by Egyptian President Gamel Abdul Nasser, in response to the withdrawal of the offer by Britain and the United States to finance the construction of the Aswan High Dam. The dam was planned to enable the Egyptian Government to control floods, provide water for irrigation and generate hydroelectricity. Britain at that time had a large military complex at Suez, which became a source of Anglo-Egyptian tension, in which the Muslim Brotherhood was a major player. In an incident in January, 1952, British attempts to disarm an auxiliary police force resulted in the deaths of 41 Egyptians, causing riots in Cairo,

and the removal of the Egyptian King Farouk. A 'Free Officers Movement' led by Muhammad Neguib and future President Gamel Abdul Nasser established an Egyptian Republic. Because of the foundation of the state of Israel in 1948, cargo shipments through the Canal had been subject to Egyptian search and possible seizure, and resulted in a Security Council resolution calling on Egypt to desist from such activities, since they contravened Article 1 of the 1888 Suez Canal Convention. Nasser's first choice as a source of weapons was the USA, but his anti-Israeli speeches made it difficult for US President Eisenhower to gain the approval of Congress to sell weapons to him. So Nasser turned to the Soviet Union, in the hope that the prospect of Egypt becoming a Soviet client State, would change the mind of the USA. In fact, Egypt obtained large quantities of Soviet arms via Czechoslovakia, to the outrage of the West, where this was seen as a major increase of Soviet influence in the region. In the Baghdad Pact of 1955, involving Pakistan, Iran, Turkey, Iraq and the UK, Britain saw the continuance of its influence in the Middle East. The conclusion of this pact coincided with an Israeli raid into the Gaza strip, in retaliation for *fedayeen* raids into Israel; this raid was commanded by *Ariel Sharon, who gave the Egyptian Army a 'bloody nose'.

Eventually, Anthony Eden was brought down by health issues. During an operation in 1963 to remove gallstones, the *bile duct was damaged, leaving him susceptible to infections of the biliary tree, biliary obstruction and liver failure. The infections took the form of ascending cholangitis, and he required more operations to deal with the situation, one of which had to be carried out in the United States, where the greatest expertise at this sort of thing lay. At this time, I was on a medical student visit to the United States, where I received much ribaldry from American medical students about Eden having to go to the United States to get matters sorted out. My response was that American surgeons were so bad that they had to develop great expertise in rectifying matters, a riposte which silenced my American colleagues. Eden had a UK country retreat at a hamlet called Binderton, which is only a couple of miles from our present home, and I am reminded of the above piece of history every time we drive past it.

The Iran-Iraq War (1980-1988)
This war concerned, amongst other issues, navigation rights in the Shatt al-Arab waterway, which is the confluence of the Rivers Tigris and Euphrates, leading into the northern end of the Persian Gulf. In 1937 Iran and Iraq signed a treaty that

settled a dispute over the control of the Shatt al-Arab. This gave Iraq control of almost all of the waterway, with the proviso that all ships using it flew the Iraqi flag, had an Iraqi pilot, and paid tolls to Iraq for each voyage through the Shatt al-Arab. However, in April 1969, Iran abrogated the 1937 treaty, and Iran therefore ceased paying tolls to Iraq for its use. The Shah of Iran, Mohammad Reza Pahlavi, claimed that most of the ships using the Shatt al-Arab were Iranian, and, therefore, the 1937 treaty was unfair to Iran. Under Saddam Hussein, Iraq's Deputy Prime Minister at the time, Iraq claimed the entire waterway up to the Iranian shore as its territory, and abrogated the treaty, despite the fact that international law holds that no treaty can be abrogated by one side alone. Iraq then invaded Iran, the main thrust of military movements on the ground being across the waterway. Since the Shatt al-Arab was Iraq's only outlet to the Persian Gulf, its shipping was severely affected by Iranian attacks, and Iraq's shipping had to be diverted to other Arab ports, such as Kuwait and Aqaba. At the end of the war, both sides agreed to once again respect the original treaty.

Saddam's actual goal was for Iraq to replace Egypt as leader of the Arab world and he gave a speech praising the Iranian Revolution. Nevertheless, he broke diplomatic relations with Iran after claiming sovereignty over some small islands in the Gulf, and squabbles over the issue of Kurdish rights.

References

1) Shaw, Karl. Royal Babylon – The Alarming History of European Royalty. Broadway Books, New York, NY 10036, USA. p.59

2) Garner J, Taylor G and Thomas A. On the origins of birds: the sequence of character acquisition in the evolution of avian flight. *Proceedings of the Royal Society.* 266, 1259-1266, 1999

Chapter 2

Metaphorical aspects

As I have explained in the Introduction, I have used the term 'metaphorical' to refer to what is commonly and widely known as 'the rat race'. Some successful individuals got to the top of their trades or professions simply because of their own intrinsic excellence. Others had to claw their way up, by fair means or foul.

The Shoguns of Japan
These first came to the attention of Europeans through Portuguese explorers, who likened the Shoguns to secular European rulers, such as the King of Spain. The first shogunate was in the Japanese Heian period (794-1185). The title of Shogun was given to military commanders who resisted the governance of the Imperial Court based in Kyoto. A warlord by the name of Minamoyo no Yoritomo established a feudal system based in Kamakura and his private militia, the samurai, became politically quite powerful, but the Emperors of Japan and the aristocracy remained the lawful rulers. Yoritomo's sons and heirs were assassinated, and the shoguns became hereditary figureheads. The Kamakura shogunate lasted from 1192-1333. There was much inter-clan strife, and the son of Emperor Go-Daigo, Prince Moriyoshi, was later killed by Ashikaga Tadayoshi, who established the Ashikaga shogunate (1336-1573), whose main base was Kyoto, which even today seems to us to be one of the most attractive cities in Japan. The Ashikaga were displaced in1603 by Tokugawa Ieyasu, who moved his headquarters to Edo, which is now known as Tokyo. The establishment of this shogunate saw the beginning of samurai control of Japan, which lasted for 700 years until the Meiji restoration in the middle of the nineteenth century. Although present day Japanese heads of government are known as 'prime ministers', when they retire they wield much behind the scenes power and are still known as *'yami shoguns'.*

Genghis Khan (1162-1227)
Genghis Khan was the founder of the Mongol Empire. He united the nomadic

tribes of northeast Asia, and proceeded to invade westwards towards Europe, his armies slaughtering all who stood in his way. His birth name was Temujin, and he was born near what is now Ulan Bator, the present capital of Mongolia. He was said to have been born with a blood clot in his hand, which, according to contemporary belief, indicated that he was destined to become a great leader. His father was subsequently poisoned by his enemies, and it was left to his mother, Hoelun, to bring up her children in the wilderness, surviving on fruits and small game. When he was 10 years old, he killed his half-brother Behter in a dispute over hunting spoils. At the age of 16 years, he was married to Börte, of the Onggirat tribe, to cement an alliance between their tribes (the Mongol Confederation). Temujin was interested in learning philosophical and moral principles from other religions, which he did by talking to Buddhist monks, Christian missionaries and Taoists. He delegated authority after considering excellence or otherwise of the candidates, rather than by being loyal to family ties. In 1200AD the main opponents of the Mongols were Naimans (in the west), Merkits (to the north), Tanguts (to the south) and Jins and Tatars to the east. He defeated and then united these tribes under his banner. At a council of Mongol chiefs, he was acknowledged as overall 'Khan' and took the name by which he is now remembered, Genghis Khan. He then went on to invade Georgia, the part of Russia around Kiev, the Crimea and Bulgaria. He was apparently castrated by a Tangut princess using a rusty knife, seeking revenge for his acts against her people. He then died, possibly by falling off his horse, but many tribes claimed responsibility for his death in one way or another, and the Tangut princess then committed suicide by drowning in the Yellow River. Genghis was, according to custom, buried without his grave being marked, but a mausoleum was created in his memory which still stands. In 1939, Kuomintang Chinese nationalist troops seized the mausoleum to protect it from the Japanese; it was disassembled and taken 900 kilometres on carts to a Buddhist monastery, the Dongshan Dafo Dian, but, owing to advancing Chinese Communist troops, it was moved again to a Tibetan monastery at Kumbum. In 1954, Genghis' bier and relics were returned to the Lord's enclosure in Mongolia, and a new temple erected to house them, In 1968, during the 'Cultural Revolution' of Mao Tse Tung, the Chinese Red Guards destroyed almost everything, but the relics were later reconstructed and an enormous marble statue of Genghis erected.

As indicated above, the Mongol Empire was based on a meritocracy. Major features of the Empire were religious tolerance, equal rights for women, and tax exemptions for teachers and doctors. I note with some disgust that the last of these did not persist into the modern era!

Attila the Hun

The year of Attila's birth is obscure, but he died in 453 AD. The Huns were a group of Eurasian nomads, who started east of the Volga and migrated into Europe in about 370 AD. They were highly warlike, their main weapons being arrows, fired from horseback, and javelins, and later, battering rams and mobile siege weapons. During their migration they attacked the Gothic kingdom between the Carpathian mountains and the Danube. During the Hun invasion, large numbers of Vandals, Alkans, Suebi and Burgundians were forced across the Rhine and invaded Roman Gaul, where they exacted tribute from the Romans, who, however, preferred to regard this as payment for services rendered. Attila announced his intention to attack the Visigoth kingdom of Toulouse, making an alliance with the Roman Emperor, Valentinian III. The latter's sister, Honoria, in order to escape a forced engagement to a Roman Senator, sent Attila her engagement ring; though Honoria may have not actually been seeking a marriage proposal, Attila interpreted her message as such an approach, and accepted, asking for half of the Western Empire as dowry! Valentinian wrote to Attila, denying the legitimacy of the alleged marriage proposal, but Attila was determined to claim what he regarded as rightfully his. Attila then gathered his vassals, and marched west, arriving in Belgica (which is now Belgium) with an army 500,000 strong. In a succession struggle concerning a Frankish ruler, Attila supported the elder son, while Aëtius, an influential Roman general, gathered troops from amongst the Franks, Burgundians and Celts. The armies clashed near Châlons, the outcome of which was a victory for the Visigothic-Roman Alliance. Theodoric the Great, King of the Ostrogoths and Germans, ruler of Italy, regent of the Visigoths and a Viceroy of the Eastern Roman Empire, was killed in the fighting.

Niccolo Machiavelli (1469-1527)

Machiavelli was a Florentine historian, politician, diplomat, philosopher and writer. His best known work is *'Il Principe'* (*'The Prince'*). He is perhaps himself most famous for his embodiment of the principle that, in order to be a successful ruler, 'the ends justify the means'. This, in modern English parlance, is therefore known as 'a Machiavellian approach' to the solution of a problem. He was born at a tumultuous time – popes fought acquisitive wars against Italian city-states, whose major players like Venice and Florence, foreign powers (France and Spain), the Holy Roman Empire and Switzerland, battled for regional influence and loot. At the age of 29, he was elected head of the second chancery of Florence, shortly after the execution of Savonarola. He carried out a number of diplomatic missions to various countries and saw the brutalities of the state-building methods of Cesare Borgia. However, matters did not proceed for him on this upward path, since the Medici, helped by Pope Julius II, used Spanish troops to defeat the Florentines, deprived him of office and accused him of conspiracy. He was arrested, thrown into gaol and tortured, but was eventually released.

He then retired to his estate and devoted himself to study and writing. In '*The Prince*', he laid down a number of principles for princely rule, including 'the end justifies the means', violence may be necessary for the stabilization of power and the introduction of new legal institutions and the justifiability of force to eliminate rivals – an approach now universally referred to as 'Machievellian'. When he died in 1527 he was buried in the Church of Santa Croce in Florence, where his gravestone epitaph reads 'TANTO NOMINI NULLUM PAR ELOGIUM', i.e. 'so great a name has no adequate praise'.

Oliver Cromwell (1599-1658)

Cromwell was the main British 'climber' in history. He was born into the gentry and was intensely religious. He was elected Member of Parliament for Huntingdon in 1628, and for Cambridge in 1628 (The 'Short' Parliament) and in 1649 to the 'Long Parliament'. In the English Civil War he was on the side of the Roundheads against the cavaliers, defenders of King Charles I, and became a principal commander of the New Model Army. He also led a campaign against the Scots between 1650 and 1651. He dismissed the Rump Parliament (see below) by force in 1653, and was later invited by his colleagues to rule as 'Lord Protector of England'. He is a highly controversial figure in English history, being considered a regicidal dictator by the historian David Hume and a military dictator by Winston Churchill, but a hero of liberty by Thomas Carlyle. In the Civil War, the conflict between Parliament and the Monarchy of Charles I, he distinguished himself at the battle of Marston Moor. A Parliamentary Ordinance of 1645 forced members of the House of Commons to choose between civil office and military command, and decreed that the army be remodelled on a national basis ('The New Model Army'). This army was commanded by Sir Thomas Fairfax, with Cromwell as Lieutenant-

General of cavalry. In 1649, at the Battle of Naseby, the New Model Army defeated the King's forces. Cromwell believed that the Army was God's chosen instrument. In 1648 those Members of Parliament who wished to continue negotiations with the King were prevented from sitting in Parliament by a troop of soldiers commanded by Colonel Thomas Pride. The remaining Members of Parliament, known as the 'Rump', agreed that Charles should be tried for treason, and Cromwell became a strong supporter of those demanding the King's trial and execution, believing that this was the only way to end the civil wars. Cromwell was very hostile to the Catholic Church, which he saw as denying the Bible in favour of papal and clerical authority and Irish Catholicism was the motive for Cromwell's Irish campaign of 1649-1650. While Cromwell was trying to negotiate a surrender, some of his troops broke into the town of Wexford, and another massacre took place, killing 2,000 Irish troops and 1,500 civilians.

In the wake of Cromwell's victories, the public practice of Catholicism was banned. Many of the principal Catholic houses of England had hiding places built both to conceal priests, and as places where the Catholic Mass could be celebrated secretly. These places were often under the roof space, in offshoots of chimneys or in ingeniously contrived cupboards, where vestments, sacred chalices and altar furniture could be stored. Many such clandestine hideouts were built by a Jesuit lay brother, Nicholas Owen, who, after the Gunpowder Plot was captured and tortured to death on the rack. He was canonised as a Catholic martyr in 1970.

In 1650 Cromwell had to leave Ireland to deal with problems in Scotland, where Charles' son, Charles II, had been proclaimed as King. He appealed to the General Assembly of the Church of Scotland, urging them to change their mind about the royal alliance, saying "I beseech you, in the bowels of Christ, think it possible you may be mistaken". Cromwell's troops were then victorious at the Battles of Dunbar and Worcester. When Cromwell returned from Scotland, he found that the Rump parliament was vacillating about setting election dates. Cromwell was so angered by this, that, supported by about 40 musketeers, he cleared the chamber and dissolved the Parliament by force. Apparently Cromwell snatched up the mace, the symbol of Parliament's power and, declaring that they were "no Parliament" and that he would put an end to their sitting, crying "take away that bauble!" In a new constitution, Cromwell was made 'Lord Protector' for life, to undertake "the chief magistracy and the administration of government". As Lord Protector, he became aware of the contribution the Jewish community had made to the economic success of Holland, England's leading commercial rival. This led him to encourage Jews to return to England; though this was the main motive, he secretly hoped that the Jews would convert to Christianity. In 1657 Cromwell was offered a crown by Parliament, but declined it, though he increasingly took on more of the trappings of monarchy. In 1658, he was struck down by a fever, possibly due to a bladder or kidney infection; he died aged 59 and was buried in Westminster Abbey. Cromwell's main legacy was the change in the government of England from an absolute monarchy to a parliamentary democracy. The concept of an absolute monarchy was derived from the doctrine of 'The Divine Right of Kings', which holds that a king is only answerable to God, and not to his people. This theory came to the fore in the reign of James I (1603-1625) of England and Louis XIV of France (1643-1715). Catholic thought justified the concept of the Divine Right firstly by reference to the Old Testament, through a prophesy of Jacob, who created his son Judah, and his descendants, to be king until the coming of the Messiah. In the New Testament, the first Pope, St Peter, commands all Christians to honour the Roman Emperor. In the Gospel of St. Matthew, Christ abjures his followers to render unto Caesar the things which are Caesar's (such as the payment of taxes), and by implication, unto God all matters spiritual.

A similar concept of Divine Right, called the 'Mandate of Heaven', existed in Asian countries. In China, there was a right of rebellion against an unjust monarch, and a despotic ruler would have his mandate withdrawn by heaven. The Zhou dynasty's rulers used this reasoning to justify their overthrow of the previous Shang dynasty. In Malaysia and Brunei, the rajahs and sultans also claimed a divine right to rule.

In the 16th century, both Catholic and Protestant politicians began to question the concept of Divine Right. The Spanish historian Juan de Mariana stated that in certain circumstances, tyrannicide could be justified. Queen Mary I was determined to restore the primacy of the Catholic faith in England, which had been overturned by King Henry VIII, when he wished to marry a Protestant, Anne Boleyn. More recent opposition to the Divine Right came from John Lock's *'Essay concerning the True Original, Extent and End of Civil Government'* and Thomas Jefferson's formulation in the United States Declaration of Independence that "all men are created equal".

Thomas Paine (1737-1809)

Paine was an English-American political activist, thinker and author. He was born in Thetford in Norfolkshire and emigrated to the American colonies in 1774 with the help of Benjamin Franklin, to whom he was introduced by a Fellow of the Royal Society and Commissioner for Excise, George Lewis Scott. Paine's principal books were '*The Rights of Man*' and '*The Age of Reason*'. He advocated colonial America's independence from Britain, and in the former book, defended the French Revolution against its critics. He was arrested and imprisoned in Paris, and released in 1794. His attacks on Edmund Burke led to a trial and conviction in his absence for seditious libel. In '*The Age of Reason*' he promotes reason and free thinking, and argues against institutionalised religion (Christianity in particular). When one contemplates the obscene death toll in wars pitting adherents of one religion against those of another, one can't help feeling that he had a point! When he died in America in 1898, only six people attended his funeral, since he had been ostracised for his ridicule of Christianity.

Arthur Wellesley, Duke of Wellington (1769-1852)

Wellesley rose to prominence during the Peninsular campaign of the Napoleonic Wars, and was promoted to Field Marshall following his victory at the Battle of Vitoria in 1813. In 1815, he commanded the allied army at the battle of Waterloo, which he nearly lost, being saved only by the timely arrival of the Prussian Army under the command of Blücher. He was twice Prime Minister, and failed to prevent the passage of the Reform Act of 1832. This legislation was largely concerned with boundary changes to the constituencies, and widening of the franchise to include small landowners, tenant farmers and shopkeepers. This was a long way from universal franchise, but a start had been made. His early military career was largely forwarded by the purchase of a majority, and then a lieutenant-colonelcy. He also fought in India. In 1802 he was promoted to Major-General, but suffered greatly from diarrhoea and trichophyton (Athlete's Foot)! He was disappointed in love, having fallen for one Kitty Pakenham, the daughter of the 2nd Baron Longford; however, he was rejected by her family, who considered him to be a poor prospect! Wellesley was a competent violinist, but burnt his instrument in anger at his rejection, and decided to pursue a military career instead. Such are the slings and arrows of outrageous fortune on which the much of the glorious history of the UK has depended! In 1793, Wellesley, having just purchased his majority, was sent to Flanders, where he bought his lieutenant-colonelcy, and had his first experience of battle. During this campaign he learnt a number of valuable lessons, including the use of steady fire lines against advancing columns of the enemy, and the merits of supporting sea-power. He said of this experience "At least, I learnt what not to do and that is always a valuable lesson". He next moved to India, where he established a list of hygiene precautions to help his men to deal with the unfamiliar climate. In 1798, the 4th Anglo-Mysore War against the local Sultan, Tipu, broke out. Here Wellesley showed what was to become one of his major attributes, namely careful logistic preparation, and he was given the additional post of chief advisor to the Nizam of Hyderabad's army, thereby exciting the jealousy of other more senior officers. This ill-will was only put to rest after the Battle of Mallavelley, in which Wellesley distinguished himself by personally leading his men in a bayonet charge, which resulted in the retreat of Tipu's forces. He became increasingly concerned at the behaviour of his men, who had taken to looting and drinking, and to restore order several soldiers were flogged and four hanged. After numerous other battles, by which time he had amassed a fortune of £42,000, he got fed up with life in India, and decided to return home, stopping, incidentally at the island of St. Helena, off the west coast of Africa, to which Napoleon was later to be exiled. When he got home. Kitty Pakenham's family withdrew their objections, no doubt because of the reputation he had gained in India, and he and Kitty were married in Dublin in April, 1806.

His next campaign was the Peninsular War (1808-1813). He arrived in Lisbon in 1809, and proceeded to rout the French commander of Napoleon's army, Marshall Soult, in Porto. In 1810, a new French army under Marshal André Masséna invaded Portugal, but it too was driven out by Wellesley's forces, and he was then promoted to full General. The French had managed to retain the Spanish fortresses of Cuidad Rodrigo and Badajoz, but were finally defeated by Wellesley's forces, amongst scenes of frightful carnage, at which Wellesley was said to have wept. After a victory at the Battle of Salamanca, and the subsequent liberation of Madrid, Wellesley was created Earl, and then Marquis of Wellington, and given command of all Allied forces in Spain. After further successes against the forces of King Joseph Bonaparte, Wellington was appointed ambassador to France.

In February, 1815, Napoleon escaped from Elba, where he had been incarcerated, and returned to France, where he regained control of the country. Wellington was sent to Belgium to take command of the British-German Army and their allied

Dutch-Belgians. The eventual showdown was, of course, at the Battle of Waterloo, on 18ᵗʰ June, 1815, which, with the help of Prussian forces under Gebhard Leberecht von Blücher (which arrived in the nick of time!), was eventually won by Wellington, after tremendous slaughter on both sides. Wellington became Prime Minister in 1828, and refused to move into 10, Downing Street, because he preferred his own home (Apsley House, at Hyde Park Corner in Knightsbridge, which became known as No. 1, London).

Amongst Wellington's political achievements was Catholic Emancipation (for which he was accused by the Earl of Winchilsea of "an insidious design for the infringement of our liberties and the introduction of Popery into every department of the State".) On this issue, Wellington challenged Winchilsea to a duel, at which both contestants deliberately fired wide of the other. However, Winchilsea did apologise to Wellington for his remarks. The latter's nickname – 'The Iron Duke' appears to have derived from this time, during which he became increasingly unpopular, and Apsley House was stormed by a mass of demonstrators, who succeeded in breaking its windows, even though the first of three Reform Bills was passed at this time, which began to shift the limited right to vote (based on property holdings) to the more universal suffrage of the present day. Wellington opposed such changes, being convinced that the country possessed a legislature which answered all the good purposes of legislation. His views did not, however, meet with the approval of public opinion, and he was forced to resign, being succeeded by Lord Grey. The Second Reform Bill was rejected by the House of Lords, triggering widespread mob violence at Derby, Nottingham, Bristol and elsewhere. A good deal of gerrymandering had been made possible by the creation of 'rotten boroughs', which were so small in population that the inhabitants could be easily bribed. For example, two such boroughs were Old Sarum in Wiltshire, which had three houses and 7 voters, and East Looe, in Cornwall, which had 167 houses and 38 voters. All this sort of business was done quite openly, with statements such as "Mr. X had been elected in Lord Y's interest", being commonly made. However, the Reform Act of 1932 disenfranchised 57 rotten boroughs, and redistributed their votes to major population centres. Rotten boroughs played a significant role in the literature of the time, e.g. in Anthony Trollope's *'Phineas Finn'* and *'Lord Silverbridge'* in the Palliser series of novels; those characters were all elected to Parliament by rotten boroughs. William Makepeace Thackeray invents the fictitious borough of 'Queen's Crawley', named in honour of a stopover made by Queen Elizabeth I, who, being so delighted by the quality of the local beer, raised Crawley to the status of a borough, giving it two members in Parliament.

Returning to Wellington, in 1833 he successfully opposed The Jewish Civil Disabilities Repeal Bill, on the grounds that England was a Christian country and had a Christian legislature, and the effect of this measure would be to remove that peculiar character. He was eventually replaced as leader of the Tories by Robert Peel, but he served in the latter's cabinets as Foreign Secretary (1834-1835) and Minister without Portfolio and Leader of the House of Lords (1841-1846). He was also reappointed as Commander-in-Chief of the British Army in 1842. His wife, Kitty, had died in 1831, and he found consolation in the company of the diarist Harriet Arbuthnot, wife of his colleague Charles Arbuthnot. When Harriet died of cholera in 1834, the two widowers lived together at Apsley House. Wellington died on 14th September 1852, aged 83, of a series of fits consequent on a stroke. One of his most famous remarks was to Queen Victoria in 1851, when the Great Exhibition was about to open, and there were a great many sparrows flying in and around the Crystal Palace. On being asked his advice on how to get rid of them, he curtly replied "Sparrowhawks, Ma'am!"

Wellington had a number of nicknames other than 'The Iron Duke'. His officers referred to him as 'The Beau', as he was an elegant dresser, and his troops called him 'Old Hookey', because of the shape of his nose. The Portuguese named him 'Douro Douro' after his crossing of the river of that name at Oporto in 1809. 'Wellington boots' refers to the substantial custom-made boots he wore instead of the standard Hessian variety. 'Sepoy General' was an insulting name used by Napoleon, referring to his service in India. 'The Beef' was yet another nickname, possibly referring to the dish 'Beef Wellington'. (in my view, this latter dish shows clearly how to ruin a nice piece of beef!)

Ludwig II, King of Bavaria – or the not so Divine Right of Kings
Ludwig was born in 1845 and became known as 'Ludwig the Mad'. His Prussian mother, Marie, was seventeen when she married, Maximilian II, King of Bavaria, of the Royal House of Wittelsbach. Most of this family were tainted by insanity, no doubt related to the degree of inbreeding amongst the Royal Houses of Europe. As an example, Otto I, King of Bavaria from 1886-1913, spent his entire reign locked up in a castle and guarded by a few medical attendants. Whilst Ludwig was engaged

in altering the Bavarian skyline by the construction of fantastic castles, and organising pan-European bank heists, Otto's madness consisted in repetitively barking like a dog, pulling faces, and taking potshots at passers-by with a rifle from his bedroom window! When it became apparent that both her sons were insane, the Wittelsbach family, looking for someone to blame, accused Marie of being responsible for the injection of defective genes into the family. Marie got increasingly concerned when Ludwig took to dressing as a nun, and complaining of strange voices in his head – a well-known manifestation of schizophrenia, though some think that this was compounded by cerebral syphilis, of which one of the well-known manifestations is delusions of grandeur. He was driven to raptures by a performance of Wagner's 'Lohengrin', and it was thought prudent to deprive him of his pet tortoise, on the grounds that he was becoming too fond of it. He was, however, God's gift to latter-day tourism, because of the fairy tale nature of his castles. He was very much of a hermit, not making a single personal appearance, since he had an aversion to breathing air also inhaled by commoners! He was also a homosexual, openly sleeping with male servants, one of whom eventually betrayed his master by giving evidence to a tribunal investigating the King's sanity. Ludwig was anxious to produce an heir, and announced that he was to marry Princess Sophie, the sister of the Austrian Empress Elizabeth, but panicked at the last moment, and got the court doctor to produce a certificate saying that he was unfit to marry. He thereafter became a confirmed misogynist and banned all females from his palace. As for Sophie, some thirty years later she strayed too near an unguarded lamp in Paris and became identifiable only by her dental chart!

Ludwig became obsessed by Richard Wagner and the fantastical background to his operas. Wagner at this time was financially in dire straits, but Ludwig rescued him from a debtors' prison. At this time Ludwig was about eighteen years old, and Wagner twice his age, and a well-known womaniser, having, amongst other liaisons, had an affair with Cosima von Bülow who was twenty four years younger than Wagner. The King settled all Wagner's debts, gave him a rent-free house and built for his exclusive use the Festival Theatre at Bayreuth. Ludwig also started building Neuschwanstein, a castle on an inaccessible peak in Bavaria. Its main subsequent use has been in films, both 'Sleeping Beauty' and 'Chitty Bang Chitty Bang Bang' being filmed there. The King regarded Bavaria's treasury as his private fund, to be raided whenever he liked! The Bavarian Cabinet gently urged Ludwig to cut down on his castle-building programme, but Ludwig ignored these pleas.

His behaviour became increasingly bizarre. He would take nocturnal rides in the Bavarian Alps, and would often seek refuge in a peasant's hut. He rode his horses around in circles throughout the nights. He ordered a servant that had failed to catch a bird that had escaped from the Royal aviary to be transported for life, and another to wear a woman's dress and ride around the palace on a donkey. He hated ugly people, and a valet who was not exactly handsome was ordered to wear a black hood whenever he was in the presence of the King "so I can't see his criminal countenance". He became obsessed with all things Chinese and made his aides wear oriental dress, and to approach him on all fours. His servants learnt to ignore his more bizarre commands. Since he was broke, he conceived the idea of selling the whole state of Bavaria, and using the proceeds to finance another castle-building programme. The Director of the Bavarian State Archives, Franz von Löher, was instructed to go abroad and find a few thousand square miles of foreign land where Ludwig could start up a new kingdom from scratch. Löher had a real foreign travel bonanza on this basis, but decided he'd pushed his luck far enough, and advised the King to stay in Bavaria. Having been refused funds by the Kings of Norway and Sweden, the Sultan of Constantinople, and the Shah of Persia, he decided to recruit thieves to break into the banks in Stuttgart, Frankfurt, Berlin and Paris. Unfortunately, he chose servants to do this who had no previous experience of bank robbery; sensibly, they went to Frankfurt, hid themselves for a few days, and then returned to explain to the King that their elaborate plans had been thwarted by unforeseen circumstances! Eventually, his subjects began to hate their King, who they suspected of spending much time buggering his cavalrymen! The King's Chief Minister, Johann von Lutz, consulted a psychiatrist, Bernhard von Gudden, asking him to diagnose the King's condition. Witnesses were found who testified that Ludwig had:-

1) Forced his soldiers to strip naked and dance for him in the moonlight.
2) Been having nocturnal picnics in the wood and eating *al fresco* in the mid-winter at temperatures well below zero.
3) Made his servants play Hunt the Slipper with him.
4) A desire to soar over the Alps in a car drawn by peacocks.
5) A pillar at Linderhof which he hugged every time he saw it.
6) Planned an exhumation of his father's body, so that he could give it a good thrashing.
7) Dispatched his Ministers to ensure that the water in his grotto at Linderhof was exactly the right shade of blue.

Von Gudden never actually examined his patient, but concluded on the above sort of evidence that the King had advanced paranoia and was incurably insane. The Bavarian government then declared that Ludwig was mad, and incapable of carrying out his duties. Von Gudden and several male nurses set out to capture him, armed with a strait-jacket and a bottle of chloroform. When they arrived at Neuschwanstein, the doors had been bolted and they were kept out by the local police. They were then arrested by the King's men, who had orders to gouge out their eyes. However, on a second attempt to capture him, the bodies of Ludwig and von Gudden were found together floating face-down in shallow water, but he exact way in which they had met their end was never ascertained.

Ludwig was succeeded by his brother Otto, who was a depressive hypochondriac and was also clearly mad. Ludwig wrote to a friend to the effect that Otto didn't take his boots off for 48 hours at a time (phew!), makes terrible faces, barks like a dog and says the most indecorous things. His doctors advised that he should be locked up for his own protection. His hallucinations became religious in content. One day, during Mass, he burst into the church, and shouted that he had sodomised several local choirboys. He died in his virtual prison in Castle Furstenried, aged sixty eight.

The descendants of the original Wittelsbach family include Prince Luitpold, who now owns the Kaltenbach brewery, which is sited in the village of Niedenweller, near Müllheim. His father, Fritz, was the third generation to run this outfit, and there was a convenient adjoining restaurant where the beer could be consumed. His son, Julius, who started by working in his father's firm, later set up on his own in the town of Haagen, and their sons, Julius and Hans took over the operation of the company.

Kings George V and VI of Great Britain
The sole real accomplishment of George V appears to have been shooting, having got the habit from his father, who presented him with his first shot gun at the age of twelve. He blasted away at pheasants, partridges, snipe, woodcock and duck, which were all easy targets because of their slow flight. On a trip to Nepal he killed large numbers of rhino, tigers and one bear. One of his fellow hunters was the Maharaja Jay Singh of Alwar, who was also a polo player. After a game which his side lost in 1933, the Maharajah poured a can of petrol over his pony and set fire

to it. King George may not have been aware that his friend was accustomed to using live babies and elderly widows as tiger bait. King George VI, his son, was similarly addicted to shooting, and included a large elephant amongst his victims. When Britain declared war on Germany, he complained that this might interfere with the grouse shouting season, which as everyone knows, starts on 12th August each year – the so-called 'glorious twelfth'. It became rather embarrassing for UK Royalty to have a German-sounding name such as Saxe-Coburg-Gotha, so Prince Philip, the husband of the present Queen, had changed his surname from Battenberg to Mountbatten.

In 1917, Tsarist Russia was in the grip of the Bolshevik revolution. The Russian Imperial Family, the Romanovs, was offered sanctuary in Britain by George V, but he had to change his mind owing to pressure of public opinion, and Lloyd George formally withdrew the offer of sanctuary. In July, 1918, the Russian Royal family were taken to Yekaterinburg and summarily shot.

The royal doctor to Kings Edward VII, Edward VIII, George V and George VI was Lord Dawson of Penn. There were said to be some doubts about his medical competence – he treated a man for jaundice for 6 weeks before realising he was Chinese!

Lenin
Vladimir Ilyich Lenin (1870-1924) was a Russian Communist revolutionary and political theorist, and was Premier of the Soviet Union from 1922 until his death. He had a major role in the October Revolution of 1917, and was responsible for the transfer of the Tsar's estates to workers 'soviets'. He concluded a peace treaty with Germany, which led to Russia's exit from the World War I. Lenin said of religion that it was a form of spiritual oppression weighing down heavily on the people, teaching them to be submissive (to their class betters) on Earth, in the hope of a heavenly reward after death – a sort of spiritual booze.

The political theory of what came to be known as Marxism-Leninism sought to purge anything considered 'bourgeois', i.e. anything suggestive of a societal class category, and anything of a religious nature, from society. "Religion is the opium of the people" is a well-known statement attributed to Karl Marx (1818-1883). Marx was born in Trier in the Russian Rhineland, and studied at the Universities of Bonn and Berlin. He moved to Paris in 1843, where he met Fredrick Engels who became his collaborator. Because of his radical writings he was thrown out and he therefore moved to London with his family in 1849. He believed that human societies progress through class struggle – between the ownership class that controls production and the 'proletariat' that provides the labour. He predicted that the internal tensions of capitalism would lead to its own collapse, and it would be replaced by a working class 'dictatorship of the proletariat', eventually leading to a classless society named 'Communism'. He advocated revolution to hurry this end. He became interested in the recently deceased German philosopher, Georg Wilhelm Friedrich Hegel (1770-1831), and his method of arguing known as 'Hegelian Dialectic'. This proceeds in three stages – a thesis, giving rise to its opposite, antithesis, and a resolution between the two, known as synthesis. Marxian dialectic is expounded in *'Das Kapital'* (1867), a four-volumed work, (to me singularly boring and not to be recommended as light reading, though, accompanied by a large whisky, an excellent cure for insomnia!). However, its world-wide impact has been enormous and has been the basis of all sorts of strife. There have been many other forms of dialectic – the ancient Greeks had two great thinkers on the subject – Plato and Socrates, who both attempted to describe the means by which we reason and argue. Socrates argued by trying to find inconsistencies in his opposer's premises and beliefs, while Plato declared that the detection of error amounts to finding proof of the antithesis. In India, in the 7th-8th century AD, philosophical debating was regarded as a spectator sport! (I don't think the modern football fan would have appreciated this!) The presiding King of the time was usually the judge at these debates, and being on the losing side could have drastic consequences, thus when a Tamil ruler defeated the Jains at debate in the 7th century AD, the punishment was the impalement of 8000 Jains! Buddhism also had institutionalised traditions of dialectics, aimed at articulating an account of the Cosmos as the truth it actually is, a system known as the Perfection of Wisdom. Karl Popper (1902-1994) has attacked dialectics for being willing to put up with contradictions. He concluded:-

"The whole development of dialectic should be a warning against the dangers inherent in philosophical system-building. It should remind us that philosophy should not be made a basis for any sort of scientific system and that philosophers should be more modest in their claims."

Popper was born in Vienna of a partly Jewish family, which converted to Lutheranism before he was born, and so he received a Lutheran baptism. He attended lectures in mathematics, physics, philosophy, psychology and the history of music at the University of Vienna. He first worked in the construction industry, but was unable to cope with the heavy labour involved. He then worked in one of the psychoanalyst Alfred Adler's clinics for children, and finally joined the University as an ordinary student. He obtained a doctorate in psychology in 1928 with a thesis entitled *'Die Methodenfrage der Denkpsychologie'* ('The question of method in cognitive psychology'). Fearing the rise of Nazism, he needed to publish something substantial to get an academic position in a country that was safe for persons of Jewish descent, and he produced a condensed version of *'Logic der Forschung'* (*'The Logic of Scientific Discovery'*). After an unpaid study visit to England he moved to New Zealand, where he wrote his most well-known and influential work *'The Open Society and its Enemies'*. He stated that no number of positive outcomes in experiments could prove a theory, but that a single counter-outcome is logically decisive and requires the theory to be discarded; falsifiability is therefore a crucial aspect of a theory, and he used this approach to debunk the claims of both psychoanalysis and Marxism to scientific status. In *'Conjectures and Refutations: the Growth of Scientific Knowledge'* he pointed out that the scientific tradition differed from its predecessors in having two 'layers'. Like the older tradition, it passed on its theories, but it also engendered a critical attitude towards them; the theories are thus transmitted down the generations, not as dogmas, but rather with the challenge to improve them.

Adolf Hitler
Possibly the most evil man in history, Adolf Schicklgruber Hitler was born in 1889 and died in 1945. He was born in Austria, and initially dreamed of becoming an artist (his few preserved paintings are actually quite good). He joined the painting school at the Vienna Academy of Art, and spent the next four years in Vienna, living off what he earned by the sale of postcards of his architectural drawings. He also started to get involved in politics and the seeds of his pan-Germanism were sown at that time. At the outbreak of World War I he enrolled in the Bavarian-German Army, and proved to be a courageous soldier, being awarded the Iron Cross (Second Class). When Hitler heard the news of Germany's defeat and the conditions of the *Treaty of Versailles (1919), he was outraged and upset, and thus was shaped the future of the world for the next several decades.

After the First World War, Hitler worked for an army organization in which his job was to check up on various political groups. During this activity he found one which he rather liked, and became its 55th member, rapidly rising to become its leader. He believed he could restore Germany to its former strength and prestige, and attempted a coup, which was unsuccessful, and for which he was imprisoned at Lansberg jail. He spent his time there writing down his ideas about a new Germany, which were later made into a book '*Mein Kampf*', ('My Struggle'). By 1932 he was in a position to run for President, but lost to Paul Hindenburg. However, shortly afterwards, Hindenburg died, and, in 1933, Hitler took over his position as President and Chancellor, combining them into one position, that of '*Führer*'. After his appointment he set about transforming Germany into the Third Reich. (The First Reich was the Holy Roman Empire of the German Nation which lasted from the coronation of Otto I as a Holy Roman Emperor in 962 AD until 1806, when it was dissolved during the Napoleonic Wars. The Second Reich was the German Empire, which lasted from the unification of Germany in 1871 until its collapse after World War I).

Back in 1919, Hitler had been appointed an intelligence agent of the *Reichswehr*, with a remit to influence other soldiers and to infiltrate The German Worker's Party (the DAP) and became attracted to the anti-Semitic, nationalist anti-capitalist and anti-Marxist ideas of its founder, Anton Drexler. Hitler was highly effective (and obviously mesmeric) when speaking to large audiences, and his meetings were supported by truckloads of party supporters waving swastika flags and distributing leaflets. A former member of the Hitler youth describes the reaction to a speech by Hitler in the following terms:-

"We erupted into a frenzy of nationalistic pride that bordered on hysteria. For minutes on end we shouted at the top of our lungs, with tears streaming down our faces:-

"*Sieg heil, sieg heil, sieg heil*'

From that moment on I belonged to Adolf Hitler, body and soul."

Early followers of Hitler included a former air force pilot, Hermann Goering, Rudolf Hess and army captain Ernst Röhm. The last became head of the Nazi's paramilitary organisation, (the SA – 'storm troopers'). On 8th November, 1923 Hitler and the SA stormed a large public meeting in a beer hall in Munich (the Bürgerbräukeller). Hitler interrupted the proceedings and announced the formation of a new government with Ludendorff. The next day Hitler and his followers marched to the Bavarian War Ministry to overthrow the Bavarian government, but he was arrested for high treason, and spent over a year in prison, where he wrote *'Mein Kampf'*. When released, he again asked Hindenburg to dissolve the Reichstag, but this building became severely damaged in a fire. The Communists were blamed, and at Hitler's insistence, Hindenburg suspended basic rights and allowed detention without trial. After an election, a new Reichstag was established, on 23rd March, 1933. An Enabling Act gave Hitler's cabinet full legislative powers for four years, and allowed deviations from the constitution. In 1934 Hitler removed the last legal remedies by which he could be deposed. This Enabling Act transformed Hitler's government into a *de facto* legal dictatorship. Hitler had already made himself Supreme Commander of the armed forces, and the oath of loyalty was altered so that it was to Hitler personally. He had by this time had his political adversaries Röhm, Strasser and the former Chancellor, Kurt von Schleicher, arrested and shot.

In 1934, Hitler appointed the Reichsbank President, Hjalmar Schacht, Minister of economics, and Plenipotentiary for War Economics. Under this regime, there was a massive fall in unemployment, as the result of a drive to improve the infrastructure of the State. Thus many roads, autobahns, railroads were constructed at this time. In 1936, the Olympic Games were held in Berlin, and Hitler employed the film director, Leni Reifenstahl, to make a film of the event, which, though obviously a piece of cinematographic virtuosity, is full of glorification of German men and women; she was another of those completely mesmerised by Hitler. (Her most famous production was *'The Triumph of the Will'*, a propaganda masterpiece). Hitler proceeded to greatly increase the size of the armed forces, against the specific provisions of the Treaty of Versailles. Thus an air force, the Luftwaffe, was formed and also a Navy, the *Kriegsmarine*, the size of which was allowed to increase to 35% of that of the British navy. As mentioned earlier. In 1936, in violation of the Treaty of Versailles, Hitler's troops reoccupied the Rhineland. In 1939, Hitler now declared Britain to be the main enemy to be defeated. He proclaimed the birth of the 'Greater Germanic Reich', in which the Dutch, Flemish and Scandinavians were to be joined in a 'racially pure' policy under German leadership. In May 1940, Germany attacked France and the Low Countries, and these victories prompted Mussolini in Italy, who always wanted to be part of the 'action', to have Italy join forces with Hitler. France surrendered on 22nd June, and Hitler's popularity rose to new heights in Germany.

Benito Mussolini (1883-1945) was a member of the Italian Socialist Party, but was expelled because of his opposition to the Party's stance on neutrality in World War I. He then formed the Italian Fascist Party, and in 1922 marched on Rome, becoming Prime Minister of Italy. In 1934, Mussolini engineered a confrontation with Ethiopia over some minor incidents in the Ogaden, concerning some wells, and the precise location of borders. In 1935, Italy attacked Ethiopia from Eritrea and Italian Somaliland, without a declaration of war, and entered Addis Ababa, thereafter announcing the annexation of Ethiopia. The League of Nations unanimously declared Italy an aggressor, but did not (or could not) take any action. Mussolini believed that following the French surrender, Italy could get territorial concessions from France, and that he could concentrate his forces on the war in North Africa, where British and Commonwealth forces were outnumbered by those of Italy. Soon after the Allied invasion of Italy in 1943, Mussolini was defeated in a vote at the Grand Council of Fascism, and the King had him arrested. However, he was rescued from prison by a commando-type raid by German forces. He then attempted to escape northwards, but was quickly recaptured and executed without trial by Italian partisans. His body was taken to Milan where it was strung up upside down for general admiration by the public.

Britain had evacuated about 330,000 troops at Dunkirk, after which Hitler made peace overtures to the new British Prime Minister, Winston Churchill, and when they were rejected, ordered air attacks on Royal Air Force bases in south-east England. The Luftwaffe was eventually rebuffed by the Royal Air Force, and when Hitler realised that air superiority could not be achieved, he ordered nightly raids on British cities, including London, Coventry and Plymouth, where we lived at the time. I was a schoolboy at Plymouth College, and every day we took various trophies we had picked up on the way to school to show our friends; these included nose cones and tail fins from incendiary bombs, and pieces of shrapnel from high explosive bomb cases. With the insouciance of youth we rather enjoyed all this. We nicknamed Mussolini 'Musso the Wop' (yes, Wop, not Wog!).

In September, 1940, a pact between Germany, Japan and Italy was signed, and later expanded to include Hungary, Romania and Bulgaria. This group became known as

'The Axis Powers'. Germany then invaded North Africa, the Balkans (including Greece and Crete, and what was then Yugoslavia) and the Middle East. In contravention of the Hitler-Stalin non-aggression pact which had been signed in 1939, millions of Axis troops attacked the Soviet Union (Operation Barbarossa), and captured huge areas of that country. They did not quite reach Moscow, since Hitler had also turned his attention to Leningrad and Kiev, and this interruption gave the Russians time to mobilise their reserves, and, when Moscow was attacked again, the German operation ended disastrously. On 7th December, Japan attacked the US naval base at Pearl Harbour in Hawaii, and four days later Hitler formally declared war on the United States. Roosevelt, in a speech to a joint session of the United States Congress, described 7th December, 1941 as "a date which will live in infamy". This was because of ongoing diplomatic negotiations with Japan, which Roosevelt characterised as having been pursued dishonestly by the Japanese Government, while it was secretly preparing for war.

After thus 'legally' gaining power, Hitler ensured the continuance of his position by putting those who disagreed with him into concentration camps. His propaganda machine, led by Goebbels (1897-1945), blamed all Germany's ills on Communists and Jews. (He had a scheme to use Jews as slaves and then kill them (about 5.5 million – the 'Holocaust') and, in furtherance of the idea of the Aryans as the master-race, he decided that they needed more room to live in *('lebensraum')*. The obvious approach was to look for it in the east, since on the west expansion was blocked by the Atlantic, the North Sea and the English Channel. His offensive into Russia ('Operation Barbarossa') was only halted at Stalingrad, probably because the lines of communication had become too stretched and because of the difficulties presented by Russian winters; the German 6th Army under Friedrich von Paulus was encircled and destroyed. Von Paulus requested permission from Hitler to surrender, but Hitler refused, demanding that the 6th Army hold its position to the last man and the last round. As a result of defeats on the Russian front, and after the D-day landings in Normandy of 6th June, 1944, many German officers decided that defeat was inevitable and would result in the complete destruction of Germany because of Hitler's misjudgements. Thus there arose a number of plots to assassinate him. One officer, Claus von Stauffenberg, planted a bomb in one of Hitler's headquarters (at Rastenburg). Hitler narrowly survived because someone had accidently pushed the briefcase containing the bomb behind a leg of the heavy conference table. Hitler ordered savage reprisals, which resulted in the execution of more than 4,000 people.

By late 1944, both the Red Army and the Western Allies were advancing into Germany, and Hitler decided to use his remaining reserves against American and British forces, which he perceived to be the weaker. He launched an offensive in the Ardennes and made some ground, causing a bulge into the Allied lines.

In his deluded state, Hitler finally concluded that Germany had forfeited the right to nationhood, and ordered the destruction of Germany's industrial infrastructure, before it could fall into Allied hands. The Minister entrusted with this 'scorched earth' policy was Albert Speer, who simply disobeyed the order. On 20th April, his 56th birthday, Hitler emerged from the bunker in which he had been hiding, and awarded Iron Crosses to boy soldiers of the Hitler Youth. He launched a tirade about the incompetence of his generals, and announced that he would stay in Berlin until the end and then shoot himself. Since Hitler was isolated in Berlin, Göring sent a telegram to Hitler, saying that he, Göring, should assume the leadership. Hitler responded by having Göring arrested. Hitler then discovered that Heinrich Himmler, head of the SS, was trying to discuss surrender terms with the Allies, and had him arrested too. On 30th April, 1945, when advancing Soviet troops were within a block or two of the Reich Chancellery, Hitler, and his long-standing mistress, Eva Braun, withdrew into a private room where in a civil ceremony the couple were married. Braun then committed suicide by biting into a cyanide capsule, and Hitler shot himself. According to Hitler's wishes their bodies were carried out of the bunker into the open, where they were thrown in to a bomb crater, doused with petrol and set alight.

Hitler's legacy was the genocidal killing of about 19.3 million civilians and prisoners of war, and, in addition, those 29 million soldiers and civilians who had died in the European theatre of action of the Second World war alone. There was no doubt that by the end of the war Hitler's health began to deteriorate, and he has being variously suggested by researchers as suffering from irritable bowel syndrome, skin problems, irregular heartbeat, coronary heart disease, Parkinson's disease, syphilis and tinnitus. Parkinson's disease would have fitted in well with the tremor of the hands shown in cinematograph films of the time, though it was clear that Hitler desperately tried to hide this. Syphilis is another possibility, since one of the complications affecting the brain is GPI (General Paralysis of the Insane), in which delusions of grandeur are often present.

In 1973, war broke out between Egypt and Syria on the one hand and Israel on the other, a part of the Arab-Israeli conflict which had been going on more or less continually since the formation of the State of Israel in 1948. During the Six Day War of 1967, Israel had captured Egypt's Sinai Peninsula and a good deal of Syria's Golan Heights, and Egypt and Syria wanted their land back. The Government of Israel voted for these returns, in return for peace agreements. Nasser was succeeded by Anwar Sadat, who declared that he would be willing to enter into a peace agreement with Israel if it withdrew from all territories occupied since 1967. Israel refused, no doubt believing that its security would be impaired. Sadat hoped that by inflicting even a limited defeat on Israel, the *status quo* could be altered. The leader of Syria, Hafez al-Assad, however, had no interest in negotiation, and started a massive military expansion, hoping that with Egypt he could then defeat Israel. Only then, he felt, would negotiations, be likely to be productive and induce Israel to give up the Gaza strip and territory on the West Bank of the River Jordan. The other Arab states were much more reluctant to commit to a new war, King Hussein of Jordan, in particular, fearing further loss of territory. Sadat was also backing the claims of Yasser Arafat, the head of the Palestine Liberation Organisation to control of the West Bank. Ariel Sharon noted from aerial photographs that the concentration of Egyptian forces along the Suez Canal was much greater than would be expected in a training exercise, and expressed concern that war was imminent.

Under such circumstances, Israel had decided it would launch a pre-emptive strike. Golda Meir, the Israeli Prime Minister, however, decided against such an approach, since she felt that American assistance would shortly be needed, and this would not be forthcoming if Israel was perceived as having started the war. On 6th October, 1973, Egypt attacked Israel by air and land, and made considerable advances. On 15th October, an Israeli counter-attack began at night, and, aided by poor Egyptian intelligence about the magnitude of the Israeli operation, Sharon pushed on, and attacked Port Said. On 22nd October, the Security Council passed a resolution calling for a ceasefire, which came into effect in the early evening. By this time, Sharon's forces were just a few hundred metres short of their goal, the last road linking Cairo and Suez. On 26th October, the Egyptian Army violated the cease fire by attempting to break through the encircling Israeli forces, but the attack was repulsed. The Golan Heights front was given priority by the Israeli High Command because of its proximity to Israeli population centres at Tiberias, Safed, Haifa and Netanya, and the Syrians were eventually ejected. There were also sea battles, which again ended in Israeli victories; throughout the war, the Israeli navy enjoyed complete command of the seas, both in the eastern Mediterranean and in the Gulf of Suez.

During all this strife, Syria ignored the Geneva Conventions and many Israeli prisoners were tortured or used as human ash trays, whilst others had their fingernails torn out, or were subjected to electric shocks and beatings. Israel had about 2,500 of its men killed inaction, and about 8,000 wounded. Arab casualties were known to be much higher, but precise figures were never disclosed. The aftermath was the Camp David Accords, resulting in the Arab-Israeli Peace Treaty. US President Jimmy Carter invited both Sadat, and the Israeli Prime Minister, Menachem Begin, to Camp David to negotiate a final peace treaty.

Margaret Thatcher (1925-2013)

Born Margaret Hilda Roberts, the daughter of a grocer in Grantham, Lincolnshire, was the longest serving British Prime Minister of the 20th century, and is still the only woman to have held that office. She was the leader of the Conservative Party from 1975-1990, and was a pretty uncompromising character, dubbed 'The Iron Lady' by a Soviet journalist. She was originally a research chemist (and thereby hangs one of her greatest gaffes – see below) before becoming a barrister, and was elected to Parliament for Finchley in 1959. Edward Heath appointed her Secretary of State for Education and Science in his 1970 government, but she defeated him in the Conservative Party Leadership election to become leader of the Opposition. She became Prime Minister after winning the 1979 General Election. Her political agenda included deregulation, especially of the private sector, flexible labour markets, and curbing the power of the trade unions. But her popularity waned during a later recession, and the accompanying high unemployment, and it was not until the Falklands War of 1982 that she regained it. She had been Head Girl at Grantham Girls' school, and had applied for a scholarship to study chemistry at Somerville College, Oxford, but was rejected and only later offered a place after another candidate withdrew. In her final year she studied X-ray crystallography under the supervision of Dorothy Hodgkin. In 1948, she had applied for a job at Imperial Chemical Industries, but was rejected after the personnel department assessed her as "headstrong, obstinate and dangerously self-opinionated". At a dinner in 1951 following her adoption as Conservative candidate for Dartford, she met her future husband, Denis Thatcher, a wealthy and divorced businessman. After being defeated at the Orpington by-election of January, 1955, she sought a 'safe' seat, and found one at Finchley. In 1967, Edward Heath appointed her to the Shadow Cabinet as Fuel and Power spokesman. When Edward Heath won the 1970 General Election, she was appointed Secretary of state for Education and Science. She certainly lacked the common touch, and when visiting a school in an impoverished area, thought she would participate in a chemistry lesson, when she said to the children "you know how a silver spoon tarnishes when exposed to the air"; however, the children had never heard of a silver spoon, let alone seen one! She ran into trouble over the provision of free school milk (which had been provided for all pupils) which she stopped, but she agreed to provide younger children with a third of a pint daily, for nutritional purposes. Cabinet papers later revealed that she had been forced into this concession by the Treasury. When all this information became public, there was a storm of protest in the Labour press, and she was dubbed "Margaret Thatcher, Milk Snatcher".

Margaret Thatcher became Prime Minister in May, 1979. When she arrived at 10 Downing Street, she got out of her car and declared:-

"Where there is discord, may we bring harmony. When there is error, may we bring truth. Where there is doubt, may we bring faith. And where there is despair, may we bring hope."

This was actually a prayer of St. Francis of Assisi, which she, somewhat surprisingly, had chosen to quote.

Her domestic policy was to reduce direct taxation and increase indirect taxes. She raised the interest rates to combat inflation and slow the growth of the money supply, and reduced expenditure on social services such as housing and education. Her cuts in spending on education resulted in her becoming the first Oxford-educated post-war Prime Minister not to be awarded an honorary degree. There were also a series of riots in England which resulted in the media discussing the need for a U-turn in policy. She replied "You turn if you want to. The Lady's not for turning", which was a pun on a play by Christopher Fry entitled 'The Lady's not for Burning'. Christopher was an inhabitant of East Dean when Barbara and I first arrived, and as with all new arrivals, he invited us to tea to inspect us. He and his wife Phyllida are buried in East Dean Churchyard, facing west down the valley, towards one of their favourite views.

In 1984, the National Coal Board proposed to close 20 of the country's 174 state-owned mines, and most of the members of the National Union of Mineworkers went on strike, led by the redoubtable Arthur Scargill despite his having lost three ballots of members. This led to the strike being declared illegal. We once met this 'bogeyman' on a train, and you could not ask for more charming and erudite company! Mrs. Thatcher declared in 1984 "We had to fight the enemy without in the Falklands. We always have to be aware of the enemy within, which is much more difficult to fight and more dangerous to liberty". Interestingly, she resisted rail privatisation, saying that it would be the Waterloo of her government.

She also confronted the Provisional Irish Republican Army (IRA) and the Irish National Liberation Army over members of those organisations being held in Northern Ireland's Maze Prison, who were on hunger strike with the objective of regaining the political prisoner status that had been granted to them by the previous Labour Government. After the death of the strikers' leader, Bobby Sands, and nine of his colleagues, some rights were restored to paramilitary prisoners, but there was still no recognition of their political status. Sinn Fein politician Danny Morrison described Thatcher as "the biggest bastard we have ever known". In November, 1981, Margaret Thatcher and the Irish Taoiseach, Garret Fitzgerald, had established the Anglo-Irish Intergovernmental Council, a forum for meetings between the two governments. Mrs. Thatcher just escaped assassination by the IRA at a Brighton Hotel in 1984, where they were staying in advance of the Conservative Party Conference. She insisted on giving her opening speech the following day, a move that increased her popularity with the public. In 1985, Thatcher and Fitzgerald signed the Hillsborough Agreement which, for the first time, gave the Republic of Ireland an advisory role in the governance of Northern Ireland.

Thatcher's next problem was the seizure by Argentina of the British colonies of South Georgia and the Falkland Islands. Argentina regarded the Falklands as with in their territory, referred to by them as the '*islas malvinas*'. There was, however, no doubt about the wishes of the Falklands population to remain British; the UK had continuously administered the islands since 1833. The economy of the Falklands is based on sheep farming, fishing, whaling, sealing, and tourism. The Argentine invasion was probably an action of the ruling military junta, led by General Galtieri, to divert attention from its poor economic performance and from growing internal opposition to the regime. On 3rd April, 1982, the United Nations Security Council issued a resolution, calling on Argentina to withdraw from the islands, and for both parties to seek a diplomatic solution. This did not happen and the UK sent an expeditionary force to retake the islands; the action was brought to a successful conclusion on 14th June, 1982 resulting in the surrender of the Argentine forces, after the deaths of 255 British and 649 Argentine soldiers, and three of the civilian population.

Thatcher was against European integration, and in a speech at Bruges, in Belgium, she opposed proposals for a federal structure and increased centralisation of decision making, she declared that "we have not successfully rolled back the frontiers of the state in Britain, only to see them re-imposed at Brussels".

Margaret Thatcher also visited China to discuss with Den Xiaoping the question of the sovereignty of Hong Kong after 1997, when the British lease of the territory was due to expire. Both governments pledged to maintain Hong Kong's stability and prosperity, and the territory was formally handed over in 1997. Unlike in the case of the Falklands, however, there were clear legal obligations on the part of the UK to hand over the territory.

Margaret Thatcher did, however, have a softer side when members of her family were in trouble. Her son, Mark, born in 1953, became lost in the Sahara Desert whilst competing in the Dakar Car Rally; his father, Denis, flew to Dakar to help organise the search, and at one time there were nine aircraft involved, from Algeria, France and Britain. The Prime Minister insisted on paying a large sum towards the cost of the search, but there was further diplomatic embarrassment when substantial unpaid hotel bills were discovered. There were media headlines about possible conflicts of interest between his business interests and his mother's political visits, including suggestions that he had received very large sums as commission in relation to an arms deal. The Prime Minister's Press secretary, Sir Bernard Ingham, suggested that he could best help the government win the next General Election, due to take place in 1987, by leaving the country! So he spent his time wandering around and living in various countries, such as Monaco, where he was granted temporary residency only, because he was on a list of 'undesirables'. He was refused residency in Switzerland and settled in Gibraltar. In 2004, he was charged with providing funding and logistical assistance in relation to an attempted coup in Equatorial Guinea. He was reported to spend most of his time in the seaside resort of Marbella. He was in

Barbados when he received news of his mother's death, and returned to the UK to act as chief mourner at her funeral in St. Paul's Cathedral on 17th April, 2013. He became 'The Honourable Mark Thatcher' following the elevation of his mother to the peerage in 1992. The courtesy title of 'The Honourable' is also shared by his twin sister, Carol.

Denis Thatcher

Denis Thatcher was a British businessman, the husband of Margaret Thatcher. He was very much a right-winger – according to himself, "an honest-to-God right winger". He was noted for his somewhat forthright comments on all sorts of matters, and was the object of spoof letters purporting to be from him to his golfing partner, Bill Deedes, a former editor of the *'The Daily Telegraph'*, in the satirical magazine *'Private Eye'*. These were written by Richard Ingrams and John Wells, and portrayed him as a reactionary, only interested in golf and gin. Wells also wrote a play entitled 'Anyone for Denis'. After ten years of Margaret's Premiership, Denis urged his wife to resign, perceiving that she would soon be forced out anyhow. He was created a baronet shortly after Margaret's resignation. He died in 2003 after a series of illnesses, including pancreatic cancer. His daughter Carol wrote his biography *'Below the Parapet – The Biography of Denis Thatcher'*, in which she recorded his admiration for American president George HW Bush, South African President FW de Klerk, King Hussein of Jordan and Mikhail Gorbachev, and his dislike for Indian Prime Minister Indira Gandhi and Sonny Ramphal, the Second Commonwealth Secretary-General. He had a soft spot for Raisa Gorbachev, Nancy Reagan and Barbara Bush. He also admitted that he wasn't quite sure where The Falkland Islands were, until the Argentinian invasion.

Mikhail Gorbachev

Margaret Thatcher was impressed by the new Soviet leader, Mikhail Gorbachev. Born in 1931 into a peasant Ukrainian-Russian family, he was initially a farmer, but graduated in law from Moscow State University in 1955. Having joined the Communist Party, he rose in 1974 to the post of First Secretary to the Supreme Soviet, and was appointed to the Politburo in 1979; he was elected to the post of General Secretary in 1985. She said of him "This is a man with whom we can do business", and, indeed, he might be said to be primarily responsible for bringing the Cold War to an end. He held summit conferences with the US President Ronald Reagan, ended the supremacy of the Communist Party in the Soviet Union, and this led eventually to the dissolution of the Soviet Union; he was awarded the Nobel Peace Prize in 1990.

Gorbachev's chief domestic aim was to revive the Soviet economy, which had become stagnant during the years when Leonid Brezhnev was General Secretary. His policy of restructuring the Soviet system ('*perestroika*') included the development of democracy, encouragement of initiative and creativity, and open criticism of failings in society and organisation. It included '*glasnost*', i.e. greater openness in expressing opinions; suppression of criticism of the government had been an important part of the Soviet system. However, he was gradually more and more opposed by one of his previous supporters, Boris Yeltsin (1931-2007). In 1987, there were two unsanctioned demonstrations in Moscow. These were strongly disapproved of by Yegor Ligachev, who upbraided Yeltsin, the effective Mayor of Moscow, for allowing them. Yeltsin wrote a letter of resignation from the Politburo to Gorbachev, who was, at the time, taking a vacation on the Black Sea. In the letter, Yeltsin expressed his frustration at the slow pace of reform, and of the servility shown to Gorbachev. In his reply, Gorbachev accused Yeltsin of political immaturity and irresponsibility, with the result that no-one on the Central Committee backed Yeltsin. After further attacks on Yeltsin, the latter tried to commit suicide. Gorbachev ordered Yeltsin from his hospital bed to attend the Moscow Party Plenum, where he was denounced and fired from his post as First Secretary of the Moscow Communist Party. Yeltsin never forgave Gorbachev for this "immoral and inhuman" treatment. Gorbachev accused him of being drunk during a visit to the USA, but these attempts to smear Yeltsin only added to his popularity. In 1990, Yeltsin was elected Chairman of the Presidium of the Supreme Soviet, despite Gorbachev's pleas to the delegates not to appoint him, and in November, 1991, he issued a decree banning all Communist Party activities in Russia. This signalled the effective break-up of the Soviet Union into a number of States, including the Ukraine, Belarus and the largest, the Russian Federation, and the Soviet Union ceased to exist on Christmas Day, 1991. There followed an era of mass privatization of State-owned industries, in which Russian citizens were given free vouchers as way of jump-starting the purchase of shares in State enterprises. However, most of the shares fell into the hands of a small number of 'oligarchs' such as Boris Beresovky, Mikhail Khordokovsky and Roman Abramovich.

Yeltsin's health was, in general, poor; he had had multiple cardiac bypass operation after a series of heart attacks, but it may be speculated that alcoholic heart disease formed part of his cardiac problem. On a stopover in Ireland, the Irish Taoiseach, Albert Reynolds was told that Yeltsin was too ill to leave the plane. On a visit to Washington, DC, Yeltsin was found on Pennsylvania Avenue, drunk, in his underwear, and trying to hail a cab! In 2005, he had a hip operation after breaking his femur whilst on holiday in Sardinia. He eventually died of heart failure on April, 2007, at the age of 76, and was the first Russian Head of State to be buried in a church ceremony since Emperor Alexander III. President Putin, the current Russian leader, declared the day of his funeral a national day of mourning. On hearing the news of his death, ex-President Mikhail Gorbachev issued a statement as follows:-

"I offer my deepest condolences to the family of a man on whose shoulders rested many great deeds for the good of the country – and serious mistakes – a tragic fate".

Sir John Major

Though many past British Prime Ministers may have been gentlemen, perhaps John Major was the only one who was obviously so. He was Foreign Secretary and Chancellor of the Exchequer in Margaret Thatcher's Cabinet and was Member of Parliament for Huntingdon from 1979 until 2001. He was Prime Minister from 1990-1997, and during his term of office he saw the rise of the European Union, a source of friction in the Conservative Party, owing to its role in the decline and fall of Margaret Thatcher. His government was responsible for the United Kingdom's exit from the European Exchange Rate Mechanism, which, in my view, might have been a good thing, since it is important that countries have full individual flexibility in financial affairs. At this time there were a series of scandals involving members of parliament and, sometimes, cabinet ministers, and Major was forced to resign. The Conservatives then lost the 1997 General Election in a landslide defeat, and made way for 'New Labour', under Tony Blair. When Major resigned, in his final statement from Downing Street he declared that "when the curtain falls, it is time to get off the stage". He then announced to the waiting Press that he was going to the Oval to watch cricket, which was one of the great loves of his life.

George Washington (1732-1799)

Washington was the first President of the United States, and the commander-in-chief of the Continental Army during the American Revolutionary War. He was born into the gentry of Colonial Virginia, and his father owned both tobacco plantations and slaves. Led by him, Revolutionary forces defeated the British Armies at Saratoga in 1777 and at Yorktown in 1781. He has been praised by historians for the selection of his senior officers, co-ordination with state governors and his meticulous attention to the questions of supplies, logistics and training. After victory in 1783, he resigned as commander-in-chief, rather than seize power *de facto*, to prove his opposition to dictatorship and he presided over the Convention that drafted the American Constitution. He established good relations with Tanacharison and other Iroquois Indian chiefs, and secured their cooperation in wars against the French, led by Joseph de Jumonville, who held large tracts of land. In 1758 he participated in an expedition to capture Fort Duquesne, but was severely embarrassed by a 'friendly fire' incident, in which his and another British Unit each thought the other was the French enemy.

In 1759, Washington had married a wealthy widow, Martha Dandridge. This was a happy union, Martha being both intelligent and gracious. His marriage greatly increased his wealth, and he became one of Virginia's wealthiest men. One of the major grievances of the American Revolutionaries was provoked by the Tea Tax, which was imposed by the British on the colonists, who believed that there should be 'no taxation without representation', meaning that they objected to being taxed by a British parliament in which they had no seats. In 1773, a group of colonists boarded the ships of the East India Company (that controlled all tea imported into the colonies) and destroyed the tea by throwing it into Boston harbour (the 'Boston Tea Party). No doubt this episode contributed to the ill-feeling against the British that led to the later rebellion that freed America from British rule.

Abraham Lincoln (1809-1865)
Lincoln was reared in a poor family on the western frontier, and became a lawyer, and a member of the Republican Party. He was a profound opponent of slavery (which formed the basis of the Southern economy), eventually pushing through the 13th Amendment to the Constitution, which permanently outlawed slavery. On his election as the 16th President of the United States in 1860, seven southern slave states immediately seceded from the Union (and seven further slave states followed) and formed a body known as the Confederate States, with Jefferson Davis as its President. The latter attacked the northern Fort Sumter on 12th April, 1861, but Lincoln's goal was to re-unite the nation. He issued a proclamation declaring the emancipation of slaves and used the army to protect escaped slaves. After the Union armies had defeated the Confederacy at the battle of Gettysburg in 1863, he addressed the crowd in the following terms which have gone down in history:-

"Four score and seven years ago, our fathers brought forward on this continent a new nation, conceived in liberty, and dedicated to the proposition that all men are created equal… It is for the living… to be dedicated here to the unfinished work which they that fought here have thus so far nobly advanced.… It is for us to highly resolve that these dead shall not have died in vain – that this nation, under God, shall have a new birth of freedom – and that government of the people, by the people, for the people shall not perish from the earth".

On 11th April, 1865, a Confederate spy, and actor, John Wilkes Booth, learnt that Lincoln would be attending Ford's Theatre, Washington, D.C. Lincoln's body guard left Lincoln's box during the interval, and Booth seized the opportunity to creep up behind Lincoln and shoot him in the back of the head, mortally wounding him. It is clearly a dangerous job being President of the United States – since in addition to Lincoln, James A. Garfield, William McKinley and John F. Kennedy suffered the same fate. There have been at least fifteen other assassination attempts on US Presidents. Of course, it is likely that the extraordinary constitutional right of Americans to bear arms cannot help! In most other allegedly civilised countries, it is an offence punishable by law! There are several notable memorials to Lincoln, notably in Washington, D.C. where a gigantic statue of a lean and scrawny Lincoln is housed in a colonnaded building and he is sculpted in granite effigies, 18 metres high, on the face of Mount Rushmore in the Black Hills of South Dakota, alongside those of George Washington Thomas Jefferson and Theodore Roosevelt.

Benjamin Franklin (1706-1790)
Franklin was one of the Founding Fathers of the USA. His father, Josiah, was a candle and soap maker, born in Northamptonshire, UK. His mother, Abiah Folger, was born in Nantucket, Massachusetts, in 1667. His father later became a successful newspaper and editor in Philadelphia, but it is for his experiments in electricity that Benjamin Franklin is perhaps best known. He was the first to suggest an experiment to prove that lightning is electricity, and flew a kite in a storm, extracting electrical sparks from a cloud. This led to his invention of the lightning rod, and used the concept of 'grounding' (i.e. earthing). For this work he received the Royal Society' Copley Medal and in 1756 he became one of the few 18th century Americans to be elected a Fellow of the British Royal Society. He was a corresponding member of the famous Lunar Society of Birmingham, which included *inter alia* James Watt, Josiah Wedgewood and Erasmus Darwin. He was also a member of the five-man committee

that drafted the Declaration of Independence. He died in 1790 from what appears to have been pleurisy and pneumonia.

Gandhi, Nehru and Jinnah

These were the leaders of the movement for independence of the various nations of the Indian Sub-continent from the British. Mohandas Gandhi (1869-1948) was commonly known as Mahatma Gandhi, or Bapu (the Father of the Nation). The word "*Mahatma*" is taken from the Sanskrit and means "great soul". He was born and raised in a community in Gujarat, on the west coast, and developed new techniques of civil disobedience to attain his ends. He led campaigns for easing poverty, women's rights and economic self-reliance. He led protests against the Salt Tax and other impositions, but always within the framework of non-violence. He declared that "There are many causes I am prepared to die for, but no causes that I am prepared to kill for" and that the concept of 'an eye for an eye' made the whole world blind". Despite his renown, he lived simply, and wore the *dhoti*, a traditional garment consisting of a piece of unstitched cloth, wrapped around the waist and legs, resembling a skirt, woven with yarn he himself had spun. This caused Winston Churchill to refer to him as "that half-naked fakir". He was a vegetarian and undertook long fasts to achieve his end, knowing that his death from starvation could not be allowed by the British for fear of the riots and violence it would cause. He was eventually assassinated by a Hindu nationalist who held him guilty of favouring Pakistan. His funeral procession was joined by over two million people. Though he was nominated five times for the Nobel Peace Prize, it was not awarded to him, but the Nobel Committee later publicly declared its regret for the omission.

After Ghandi's death, Nehru and Patel, the two most influential people in the Indian Congress, agreed that the first priority must be to stem the hysteria and likely violence. They called on Indians to honour Gandhi's memory and observe his ideals. At this time there was great tension between the Hindu and Muslim communities, so the government had to make sure that everyone knew that the assassin was not a Muslim. Gandhi was cremated, the usual practice, and his ashes widely distributed in urns sent across India to be present at memorial services.

Jaharwarlal Nehru (1889-1964)

Nehru was the first Prime Minister of independent India, the father of Indira Gandhi and the grandfather of Rajiv Gandhi, who all became Prime Ministers in due course.

(It should be noted that 'Gandhi' is a very common Indian name, and having it does not imply being related to the Mahatma). He was a the principal author of the Indian Declaration of Independence of 1929. He had been imprisoned by the British, and, when he emerged, he found that the political landscape was much changed. His old Congress colleague, Muhammad Ali Jinnah (1876-1948) was the founder of Pakistan. He was born in Karachi, and trained as a lawyer at Lincoln's Inn in London.

Muhammed Ali Jinnah 1876-1948)
Jinnah was leader of the All-India Muslim League from 1913 until Pakistan's independence, in August 1947, and then Pakistan's Governor-General until his death. He was concerned to maintain the political rights of Muslims in India. However, he resigned from the Congress Party when it agreed to follow the policy of non-violent resistance advocated by Gandhi. By 1940 he had come to believe that Indian Muslims should have their own state, and this led the Muslim League to pass a resolution in Lahore demanding a separate nation. All parties agreed to work towards a separate independence – for a predominantly Hindu India, and for a predominantly Muslim state, which was to be called Pakistan. When Jinnah began his legal practice in Bombay (now Mumbai), he was the only Muslim barrister in the city. One of Jinnah's fellow barristers in the Bombay High Court commentated on Jinnah's supreme faith in himself, and recalled being told off by a judge in the following terms:-

"Mr. Jinnah, remember that you are not addressing a third-class magistrate!"

Jinnah's riposte was "My lord, allow me to warn you that you are not addressing a third-class pleader!"

Relations with Britain became strained when the Imperial Legislative Council promulgated emergency restrictions on civil liberties, and Jinnah resigned. There was general unrest in India which got worse after the Amritsar massacre. This took place when a British officer, Brigadier-General Reginald Dyer, became convinced that a major insurrection was about to take place, and banned all meetings. Nevertheless something between 15,000 and 20,000 people, including women, children and the elderly gathered; Dyer, arriving with a group of Gurkha and Baluchi soldiers, believed that a violent thrashing would dampen any conspiracies afoot, and, without warning the crowd to disperse, blocked the main exits from the square where the meeting was taking place. He then ordered his troops to shoot into the densest parts of the crowd, and the shooting continued for ten minutes. Although the precise figures are disputed, it is likely that between 1,000 and 1,500 were killed. Information about the massacre was suppressed in Britain, but the news spread like wildfire in India. The widely respected poet and author, Rabindranath Tagore, renounced his British knighthood as a symbolic act of protest. In October, 1919, however, the Secretary of State for India, Edwin Montagu, set up a committee of inquiry, comprising both Indian and British members, with the remit of investigating the disturbances in Bombay, Delhi and Punjab. It found that:-

1) Lack of notice being given for the crowd to disperse was an error.
2) The duration of firing was unnecessarily long.
3) Dyer's motive of producing a 'sufficient moral effect' was to be condemned.
4) Dyer had overstepped the bounds of his authority.
5) There had not been any conspiracy to overthrow British rule in the Punjab.

In a minority report, the Indian members added that the proclamations banning public meetings were insufficiently widely distributed, that there were innocent people in the crowd and that there had not been any previous violence. They said that Dyer should have either ordered his troops to help the wounded, or instructed the civil authorities to do so, and his actions had been inhuman and un-British and had greatly injured the image of British rule in India. The Viceroy's Council found that Dyer was guilty of a mistaken notion of duty, and he was relieved of his command! He had been previously been recommended for the award of Commander of the British Empire, but this was cancelled. There were other aftermaths, e.g. an Indian independence activist, Udam Singh, shot and killed Michael O'Dwyer, the British Lieutenant-Governor of Punjab at the time of the massacre and he was hanged for the murder. At the time, Nehru and Gandhi condemned his action as "useless but courageous". In 1952, Nehru, by then Prime Minister, saluted Singh "who had kissed the noose so that we may be free". Years later, Queen Elizabeth II visited Amritsar and paid her respects with a half-minute period of silence. In February, 2103, David Cameron, the British Prime Minister, laid a wreath at the memorial in Amritsar, and described the massacre as a deeply shameful event, one that Winston Churchill rightly described at that time as "monstrous." The event has already been the basis of a number of films, including '*Gandhi*' and '*The Jewel in the Crown*'. In 1981, Salman Rushdie in his novel '*Midnight's Children*' portrayed the massacre as in the eyes of a doctor in the crowd. There are also two films in Hindi which centre on the massacre.

The Emperor Hirohito of Japan

Hirohito was the first son of Crown Prince Yoshihito (the future Emperor Taisho and Crown Princess Sadako (the future Empress Teimei). He had military training in the army and navy) and married a distant cousin, Princess Nagako Kuni in 1924. He succeeded to the throne on the death of his father, Yoshihito, in 1926. He survived an assassination attempt by a Korean independence activist in 1932. Japan invaded Manchuria and parts of China in 1937 because Hirohito feared an attack by the Soviet Union. He personally ratified his Army's proposal to remove the constraints of international law so far as the treatment of Chinese prisoners was concerned and he also sanctioned the use of chemical weapons. In 1940, Japan formed a Tripartite Pact with Nazi Germany and Fascist Italy, forming the bloc known as the 'Axis Powers'. At an Imperial Conference at this time, all speakers were in favour of war rather than diplomacy. Hirohito chose the hard-line General Hideki Tojo as his military leader, and asked him to provide provocation for war. The US Secretary of State for war, Henry L. Stimson recorded in his diary that he had discussed with the US President (Franklin Roosevelt) the likelihood that Japan was about to launch a surprise attack, and on 8th December, 1941, Japanese forces struck at the US base in Hawaii at Pearl Harbour, and also at the Philippines and at what was then Malaya. At first, all was success for the Japanese. But the tide began to turn at the battle of Midway, and with the landing of US forces on Guadalcanal and Tulagi in August, 1942. In June, 1944, Hirohito sent out an order encouraging all Japanese civilians to commit suicide rather than be taken prisoner. (Suicide was an ingrained part of Japanese culture – *seppuko* or *hara-kiri*).

Japan was doing basic research on the atomic bomb, but, as everyone knows, the Allies got there first, and on 15th August, 1945 Japan surrendered. Hirohito declared:-

"Moreover, the enemy has begun to employ a new and most cruel bomb, the power of which to do damage is, indeed, incalculable, taking the toll of many innocent lives. Should we continue to fight, not only would it result in an ultimate collapse and obliteration of the Japanese, but it would also lead to the total extinction of human civilization". As the situation for Japan had become increasingly desperate, the Japanese resorted to the use of *kamikaze* pilots. *Kamikaze* aircraft were basically used as pilot-guided explosive missiles, and the pilots would attempt to crash their planes into enemy ships (particularly aircraft carriers); at least 47 Allied vessels were destroyed in this way. Obviously, accuracy was much better than that achieved than by a conventional bombing attack. When wondering at the self-sacrificing bravery of these pilots, it should be noted that the tradition of death, rather than the shame of defeat and capture, was deeply engrained in Japanese culture, and was part of the *samurai* life and the *bushido* code of loyalty and honour until death.

After the end of World War II, there were attempts to try Hirohito for war crimes, such as the gross mistreatment of prisoners and civilians. However, US General Douglas MacArthur insisted that Hirohito retained his throne. Indeed, there was a successful campaign to absolve the Emperor from all responsibility for war crimes, so much so that he was turned into an almost saintly figure who bore no moral responsibility for the war. The worse that happened to him was that he was forced to renounce the idea that the Emperor was an incarnate divinity. For the rest of his life, Hirohito remained an active figure in Japanese life, performing the functions usually associated with a constitutional monarch. He travelled abroad and met Queen Elizabeth II and US President Gerald Ford. He was also heavily involved in marine biology and respected for having published several papers under his own name. In 1987 he underwent pancreatic surgery, and his doctors discovered that he had duodenal cancer (though this doesn't sound right to me, since cancer of the duodenum is extremely rare). His state funeral was held on 24th February, 1989.

Finally, readers may wonder at my natural reticence in not including an account of my own rise to my present modest position. This is mainly because I have already written something about it in my book 'CORPUS HOMINIS – memoirs of an academic physician'. Here I will complement that ' masterpiece' by making a number of negative assertions:-

1) I have never slept with the wives of influential people
2) I have never shot the dogs of my neighbours, though I would have dearly liked to – there always seems to have been someone looking on whenever I pulled out my AK47.
3) I have never driven a stake through the evil heart of a vampire, nor drank the blood of even one such creature.
4) I have never attempted to bribe a judge.

5) I am not in the habit of deflowering virgins.

Ladders in music

Perhaps the most obvious example in music of clawing one's way to the top of the pile comes in an opera by Modest Mussorgsky (1839-1881), *'Boris Godunov'*. The historical background to the opera, relates to the *de facto* regent for Ivan the Terrible's feeble son Fyodor, who cared only for spiritual matters. When Fyodor died, the Patriarch Job of Moscow nominated Godunov to succeed him, despite rumours that Boris Godunov had ordered the assassination of the eight year-old Tsarevich Dmitry Ivanovich. In the opera, Boris then sings his famous aria "I have obtained the highest power". One of Mussorgsky's other well-known works is 'Pictures at an Exhibition', which describes an imaginary tour of an art collection, and is a favourite of virtuoso pianists As the viewers walk from picture to picture, there is a piece of interlude music, known as the 'promenade'. Some of the pictures depicted in this music are *'Two Jews, rich and poor'*, *'Samuel Goldenberg and Schmuyle'*, *'Gnomus'*, *'Ballet of the Unhatched Chicks'*, *'Catacombs'* and *'Baba Yaga'*.

Prevention

It is sometimes necessary to prevent something or someone from moving either up or down. Perhaps the best guide to these skills is provided by Cornford, who describes the Technique of the Alternative Proposal, and is used at committees to torpedo propositions one doesn't like! Someone wanting to use this technique arranges that several members of the committee will make different proposals for the the way forward on whatever is being discussed. As soon as three or more alternatives are in the field, there is almost certain to be a majority against any one of them, and nothing will be done"!

If you belong to a College, or other academic institution, an appeal to College feeling may work. This relies on your heartfelt belief that your institution is better than any institution to which anyone else belongs, thus promoting a state of healthy rivalry. This feeling makes the college system very valuable, and distinguishes it from a Boarding House, in which hatred is usually concentrated, not on rival institutions, but on other members of the same establishment.

Cornford prefaces his advice with the following verses:-

"If you are young, do not read this book; it is not fit for you;

If you are old, throw it away; you have nothing to learn from it;

If you are unambitious, light the fire with it; you do not need its guidance.

But, if you are neither less than twenty-five years old, nor more than thirty;

And if you are ambitious withal, and your spirit hankers after academic politics;

Read, and may your soul (if you have a soul) find mercy!'

There can be few retired academics for which these lines do not stir some sort of memories!

Reference

1) Cornford, FM. *Microcosmographia Academica* – being a guide for the young academic politician. Bowes and Bowes, Cambridge, 1908

Notes

Anti-hypertensives
These are drugs, of many types, which are used to lower blood pressure.

Apoproteins
Fats are not soluble in the liquid phase of plasma and therefore circulate attach to 'apoproteins' which are suspended in the plasma.

Atherogenesis
The process by which arteries become lined with cholesterol, phospholipid and fatty plaques, which may eventually block the artery and cause critical narrowing, thrombosis or the formation of clots in the narrowed areas.

Bile duct
This is the passage conducting bile from the liver to the duodenum. Bile is an essential ingredient of the digestive process.

Crown of Thorns
In the present context, this is a large multiple-armed starfish widespread in the Indian and Pacific Oceans which destroys coral reefs. Stepping on one is highly painful, and the spines contain a saponin-based toxin.

Gallipoli Campaign
By the end of this episode in World War I, over 100,000 men had died, including about 60.000 Turkish and 53,000 British and French soldiers, 8,709 Australians and 2,721 New Zealanders. In total, there were nearly 500,000 casualties.

Geisha
A geisha is a traditional Japanese hostess, whose skills include performing various Japanese arts, such as classical music, dance, games and serving at ceremonial dinners. There are complicated geisha ranking systems, the highest (and most expensive)

being patronised by businessmen and politicians. They always wear kimonos, with an extravagant '*obi*', a coloured sash around the kimono. They sport extravagant hairstyles and always sleep with their necks on small supports, to avoid disturbing their hair. They are always single women and, if they wish to marry, they must retire as geishas. Though in the late 7th century AD some might have been prostitutes, by the 8th century, when the imperial court moved the capital to Kyoto, they had become a beauty-obsessed elite, with extraordinary complex systems of making themselves up, featuring a white base, with red lipstick and red and black accents around the eyes and eyebrows; originally the white base contained lead but when the toxicity of lead began to cause skin problems, it was replaced by rice powder. Japanese men were not constrained to be faithful to their wives, the ideal wife was a mother and manager of the home, and for sexual enjoyment, men went elsewhere, to courtesans. World War II caused a great decline in the geisha arts, since most women had to go to factories to contribute in one way or another to the war effort.

Gorbachev, Raisa

Mikhail Gorbachev's wife, Raisa, became well-known to the West during her travels with her husband. Her public appearances as first lady were a novelty at home in Russia, and went a long way to humanising the Soviet image. She raised funds for the preservation of the Russian cultural heritage, the fostering of new talent, and for treatment programs for children's cancer.

Low-density lipoprotein

This is a circulating protein, and is the main carrier of cholesterol in the bloodstream.

Mossad

Mossad is the national intelligence agency of Israel. The actual word itself is the Hebrew for 'The Institute'. It is responsible for intelligence collection, covert operations and counter-terrorism, and for bringing Jews to Israel from countries where official 'Aliyah' agencies are forbidden. ('Aliyah' is the Hebrew for returning to Israel from the Diaspora, i.e. from far-flung countries of the world in which Jews had to take refuge in harder times). Its most famous operation was the capture of Adolf Eichmann, and bringing him to Israel for trial.

Osteonecrosis

Death of bone

Perspective in art
This is the technique by which paintings on a two-dimensional surface are made to appear three-dimensional.

Radiation sickness
This is the syndrome caused by exposure to high levels of ionising radiation; the problems include nausea, vomiting and problems related to destruction of the highly radiation-sensitive bone marrow, and the later development of various cancers.

Roosevelt, Franklin Delano (1882-1945)
Franklin Roosevelt (FDR) was the 32nd President of the United States from 1933 till 1945, and the only one ever to serve three terms. He rose to be leader of the Democratic Party, and built a New Deal Coalition in response to the Great Depression, which was focussed on relief for the unemployed, reform of the economy so as to regain previous levels of activity, and of the financial system to prevent a repetition of the Depression which had devastated the nation in the 1930's. As Roosevelt took the oath of office on 4th March, 1933, the Governors of the individual States had closed every bank in the nation, and no-one could cash a cheque or withdraw their savings. This Depression only ended when the USA entered World War II in December 1942 in the aftermath of the attack on Pearl Harbour.

Sharon, Ariel
Sharon was an Israeli statesman and former general, considered the greatest field commander in Israel's history. He was born in 1928, and at the age of 10 years joined the Zionist youth movement Hassadeh. At the age of 14 he joined the Haganah (an underground paramilitary force and the precursor to the Israel defence forces). He eventually became commander of a Special Forces Unit, formed to take reprisals against the *fedayeen*. He had led an armoured division in the Yom Kippur War. He became Israel's 11th Prime Minister. He suffered from numerous health problems, including obesity, high blood pressure and high blood cholesterol. His staff car was stuffed with cigars, snacks, vodka and caviar. In December, 2005, he suffered the first of a series of strokes which left him in a permanent vegetative state, i.e. alive, in a state of partial awareness, but unable to communicate and died aged 85 after eight years in a coma. He was succeeded as Prime Minister by Ehud Olmert in 2014.

Treaty of Versailles
This treaty, signed in 1919 in the Hall of Mirrors at the Palace of Versailles, required

Germany to accept the responsibility for causing all the damage and loss of life in the First World War, and provided for the trial of war criminals. It imposed strict limitations on the size and content of the armed forces permitted to Germany, and, in particular, prohibited the use of poison gas in future wars after chlorine had been deployed by the German Army north of Ypres; chlorine is a powerful irritant to the eyes, nose, throat and lungs, probably working through the production of hydrochloric acid. However, it is water-soluble, and therefore breathing through a wet cloth somewhat reduces its damaging effects. Its use is dependent on the wind being in the right direction, and if that direction is variable, then its deployers may get a taste of their own medicine, as happened to the British at the Battle of Loos in 1915. This trouble was increased when it was found that the wrong keys for opening the gas canisters had been supplied. Phosgene was a more potent agent, but with the drawback that it took 24 hours or more to take effect. This meant that the victims could carry on fighting until the effects took hold. It was at the battle of Loos that the gas mask was used for the first time. The British version contained sodium hyposulphite, which was partially effective against phosgene (see below), and was worn over the top of the steel helmet until a gas alarm was given, when it was pulled down over the face. I remember well the various designs of gas mask which we had to carry about during World War II, there being versions suitable for babies, children and adults. The Scottish Highlanders were said to be especially vulnerable to mustard gas, which is a painful skin irritant, because they wore their kilts into battle and thus had partially bare legs. Poison gas may be delivered, as by the Germans in the case of mustard gas, in artillery shells, thereby avoiding much of the trouble from changes in wind direction and a dense concentration of gas could be effected by firing them from large batteries of mortars. The Germans later used Zyklon B, a cyanide-based gas as an instrument of extermination in their concentration camps

Phosgene has the chemical formula CCl_2O, and was stockpiled by some countries as part of their armamentarium. It was used by the Imperial Japanese Army in the Second Sino-Japanese War, its use being specifically sanctioned by Emperor Hirohito. Its action is to inflame the membranes lining the alveoli of the lungs, causing suffocation. There was also some limited protection given by the various types of gas helmets developed at the time. Gas was seldom employed in World War II, but it was used in the Iran-Iraq War of 1980–1988, the result of border disputes between the two countries.

Index

Abramovich, Roman 104
Adam, Adolphe 55
Abu Dhabi 6
Aerofoil 25
Afrika Corps 42
Airships 19, 20
Al-Assad, Hafez 98
Altensalzkoth 45
Altitude sickness 32
American Constitution 70, 106, 107
Ampikaipakan 53
Amritsar massacre 109
Anschluss 44, 66
Anti-aircraft guns 25, 37
Antwerp 36
Apollo 11 mission 39
Apsley House 85, 86
Arafat, Yasser 60, 98
Archaeopterix 13, 14
Argentina 45, 46, 101, 102
Arno, River 16
Asbestos 3
Attila the Hun 9
Attlee, Clement 41, 42, 67
Auchinleck, Claude 42
Aviation 13, 18, 19, 23, 24
Avon Gorge 48
Axis Powers 96, 11
Aylieff, Felicity 10
Azores 6

Bad Frankenhausen 6
Ballistic missiles 35, 36
Baluchis 109
Barrage balloon 25, 37
Bazooka 35
Beatles 10, 11
Beckenham, Kent 5, 49
Beethoven, Ludwig van 38
Beggar's Bush 48
Begin, Menachem 99
Belarus 104
Bénezet 55
Ben-Gurion, David 46
Beresovsky, Boris 104
Bevan, Aneurin 41, 42, 52, 53
Big soups 2
Bing Joseph 86
Biodiesel 29
Bisphosphoglycerate 2,3 32
Blair, Tony 105
Blanchard, J-P 18
Blériot, Louis 22
Blimps 19
Blücher, Gebhard 83, 85
Boeing 747 23
Boleyn, Anne 82
Bombay High Court 109
Booth, John Wilkes 107
Boris Godunov 113
Boston Tea party 106

Boucher, Barbara J. i
Bourdillon, Tom 33
Bramante, Donato 12, 17
Braun, Wernher von 37, 38
Brezhnev, Leonid 73, 104
Bridges 46, 47, 49, 50, 53, 5-57,
    59, vii
  Alcantaro 46
  Arkidako 46
  Avignon 45
  Chinese 46
  Clapper bridges 47
  Clifton Suspension 48, 49
  Eades 40
  Golden Gate 51, 53
  Ironbridge 47
  Millennium Bridge 51
  Penang Bridge 49, 53
  Sydney Harbour 53
  Tagus, River 46
  Tiber 46
  Zhaozhou 46
Brunel, Isambard Kingdom 49,
    51, 52
Buchaille Etive Mor 31
Buddhism 34, 91
Buenos Aires 45
Bungee jump 49
Burke, Edmund 48, 83
Boris Godunov 113
Bonaparte, Napoleon 84
Bush, George 103
  Barbara 103
  George H. 103
Bushido 112
Byzantine style 16

Cameron, David 110
Camp David Accords 99
Carrel, J-A 31
Carter, President Jimmy 99
Catesby, Robert 34
Catholic Emancipation 107
Catholicism 15, 34, 81, 82
Castles 5, 22, 54, 86-89,
Cavendish, Henry 18
Cerebral oedema 32
Certon, Pierre 55
Chagall, Marc 11, 17
'Challenger' disaster 39
Charlemagne 38
Chichester 7
Chrysostom, St. John 9
Churchill, Winston 37, 43, 44,
    65, 65-69, 80, 95, 108, 110
Cimabue 16, 17
Clapper Bridge 47
Clifton College, Bristol 48, i
Climacus, St. John 10
Clock Face Pit 41
Cohen, Robert D 3
Concorde 23, 24, 49
Confederate States 61, 106
Congress Party, Indian 109
Cornford, FM 113, 115
Crinoline 48
Cromwell, Oliver 80, 81, 82
Croz, Michel 31
D-Day landings 36, 38, 96
Dachau 44
Dakar Car Rally 102
Dandridge, Martha 106
Dartmoor 47, 48, 49

Darwin, Erasmus 107
Davis, Jefferson 82, 106
Day of Atonement 15
Decompression sickness (DCS) 40
Deedes, Bill 103
Declaration of
   Independence, America 82,
   108, 109
Den Xiaoping 102
Devonport 50
Dhahran 36
Dhoti 108
Divine Right of Kings 11, 82, 86
Doenitz, Karl 63
Downing Street 4, 85, 100, 106
Douglas, Lord Francis 31
Dover Castle 22
Dresden 17
Dürer, Albrecht 10, 17
Dyer, Brigadier-General R 109,
   110
Eads Bridge, Missouri 40
East Dean, West Sussex 35, 100
East India Company 106
Eicosapentenoic acid 28
Eiger, North Face of 33
Eisenhower, Dwight D. 57, 74
El Alamein 68
Equatorial Guinea 102
Erythropoetin 32
Essential fatty acids 28
Euphrates, River 74
European Exchange
   Rate Mechanism 105
Evans, Charles 33
Fakir 43, 108

Falklands Islands 62, 99, 101,
   102, 103
Fawkes, Guy 34, 35
Final Solution, The 45
Filton (airport) 49
Fireworks 38, 74
Fitzgerald, Garret 101
Florence 16, 80
Ford, Gerald 112
Fort Duquesne 106
Franklin, Benjamin 83, 107
Fry, Christopher 100
Galileo 6
Galtieri, General 101
Gandhi, Indira 103, 108
Gandhi, Mohandas (Mahatma)
   43. 44, 107, 108
Garfield, James A. 107
Garrick, David 48
Garter, Order of 11
Gears 26
   Sturmey-Archer 26
   Derailleur 26, 27
Geishas 118
Geneva Protocol 1925 56, 99
Gettysburg, Battle of 106
Gurkha 109
Giordano, Felice 31
Giotto 16, 17
'Glasnost' 73, 104
Glen Shiel 4
Goering, Hermann 23, 93
Gogh, Vincent van 10
Golan Heights 98
Goldsmith, Oliver 48
Goodwood race-course 25, 30

Gorbachev 104
    Mikhail 73, 103, 105
    Raisa 118
Grantham Girls' School 99
Gravity, Einstein's Theory of 7, 21
'Grecian bends' 40
Guadalcanal 111
Gujarat 108
Habeler, Peter 33
Haemoglobin/oxygen
    dissociation curve 32, 34
Haifa 44, 98
Hamoaze 49
Hara-kiri 111
Hawker Tempest 37
Heath, Edward 99, 100
Heinz, HJ 2
Helium 19, 20
Heinkel 112 38
Heydrich, Reinhard 45
Hillary, Edmund 33
Hillsborough Agreement 101
Himalayas 32, 34
Himmler, Heinrich 38, 45, 97
Hindemith, Paul 38
Hindenburg 20
Hirohito 111, 112, 120
Hitler, Adolf 38, 42, 46, 56, 64-
    69, 71, 73, 92-97
Hodgkin, Dorothy 99
Home Guard 25
Hong Kong 102
Hooke, Robert 8, 86
Hudson, Charles 31
Hunt, Sir John 33
Hussein, King, of Jordan 98, 103

Hydrogen 18, 19, 20, 38
Incas 46
Ingham, Bernard 99
Iran 74, 75
Iran-Iraq war 36, 74
Iraq 74, 75
Irish National Liberation Front
    60, 98
Irish Republican Army 4
    (IRA), Provisional 101
Iron Bridge, Telford 47
Iroquois 106
Jacobite Rising 4
Jefferson, Thomas 82, 107
Jeffries, J 18
Jet lag 24
Jinnah, Muhammed Ali 108, 109
Johnson, Samuel 48
Kamikaze pilots 111
Kenaf 29
Keratin 13
Klerk, FW de 103
Ladders iii, 1-11. 13, 15-16, 17-
    21, 30
    assault 3
    in stockings 8
    Jacob's 9, 10
    kitchen 2, 7
    orchard 5
    roof 1, 4, 5
    series 1
        convergent 1
        divergent 1
        infinite 1
    walking under 8, 9
Kaltenbrunner 44, 45

Kennedy, John F. 107
Kettlewell 30
Khordokovsky, Mikhail 104
Kilnsey Crag 30
King's House Hotel, Glencoe 31
L'Auberge Pleine 55
Led Zeppelin 20
Leonardo da Vinci 17
Lhasa 34
Lhotse 33
Library steps 7
Lift equation 21
Ligachev, Yegor 104
Lighthouses 5
Lincoln, Abraham 106, 107
Lindbergh, Charles A. 23
Linz 44
Lloyd George, David 42, 6-65, 90
Lock, John 82
Lord Protector 80, 81
Lord Warden of the Cinque Ports 22
Louis XVI 18
Louis X1V 82
Ludwig II, King of Bavaria 86-89
Lunar Society of Birmingham 107
Lüneberg Heath 45
Macaulay, Thomas Babington 47
Magna Carta 22
Mahathir Muhammed 53
Major, John 4, 105
Makalu 33
Malaya 111
Mallory, George 32, 33
Manetti 16
Marbella 102

Marie Antoinette, Queen 18, 38
Matterhorn 31, 33
McAuliffe, Christa 39
McKinley, William 107
Maze Prison 101
Meir, Golda 99
Messerschmidt ME 202 36
Messner, Reinhold 33
Michelangelo 12
Microcosmographica academica 115
Montagu, Edwin 110
Montgolfier brothers 18
Montgomery, Field Marshal Bernard 57
Morris, William 41
Morrison, Danny 101
Mortars 4, 120
Mossad 45, 46, 118
Mussorgsky, Modest 113
Mustangs (P-51) 37
National Coal Board 101
National Health Service 42, 52
National Union of Mineworkers 101
Nazi Party 23, 38, 39, 44-46, 62, 69, 93
Nazism 23, 92
Nehru, Jawaharlal 108, 110
Nepomuk, St. John of 54, 55
Netanya 48
Neuschwanstein Castle 87, 89
New York 23, 24, 51, 52, 61
Newton, Isaac 7, 8
Nobel Peace Prize 104, 108
Nuclear reactors 25, 62

Nuptse 33
O'Dwyer 110
Oligarchs 104
Operation Desert Storm 36
Osteonecrosis 40, 118
Ostrogoths 79
Owen, Hugh 34
Paine, Thomas 83
Pakistan 76, 108, 109
Palestine 44, 60, 64, 98
Palestine Liberation Organisation
    60, 98
Pan-American Airways 23
Parsonage Pit 41
Patriot missile 35, 36
Patton, George S. 57
Pavia, Civic Tower of 6
Pearl Harbour 23
Pearly Gates 15
Peel, Robert 86
Penang bridge 49, 53
Peninsular War (1808-1813) 83,
    84
Penn, Lord Dawson of 90
'perestroika' 104
Persian Gulf War 36
Philadelphia 24, 107
Philippines 111
'Pictures at an Exhibition' 113
Piero della Francesca 11, 17
pH, of blood 32
Pisa, Leaning Tower of 5
Pitti Palace 16
Plympton 48
Polycythaemia 32
Pope Julius II 12, 80

Pope Pius V 16
Porsena, Lars, of Clusium 46
Postbridge 47
Pre-Raphaelite Brotherhood 41
'Private Eye' 103
Pulley systems 29
Pulmonary oedema 32
Putin, Vladimir 105
Queen Mary I 82
Ramphal, Sonny 103
Reagan 39
    Nancy 103
    Ronald 39, 103
Reynolds, Albert 105
Reynolds, Sir Joshua 48
Rhône, River 55
Richard the Lionheart 11
Rivers 47, 49, 53, 74, 86
    Douro 86
    Lynher 49
    Tamar 49
    Tavy 49
    Tiber 48
Rockets 34, 35, 37-39
Reform Bills 83, 85
Romanovs, The 90
Rommel, Erwin 42
Roosevelt, Franklin D. 68-70, 96,
    111, 119
Roosevelt, Theodore 107
Ropes 1,6,16,27,29,30
    Coir 27
    Cotton 27
    Flax 27
    Hemp 27
    Jute 27

Linen 27
Sisal 27
Straw 27
Rotherhithe Tunnel 49, 51
Rotten boroughs 85
Royal Ulster Constabulary 4
Rugby football 48
Rump Parliament 80, 81
Rushdie, Salman 110
Rushmore, Mount 107
Russian Federation, The 104
Sadat, Anwar 98
Santa Croce, Church of 16, 80
Santa Maria Novella, Church of
    16
St. Francis of Assisi 100
St. Peter's Basilica, Rome 12
Saddam Hussein 75
Safed 98
Salt Tax 108
Saltash Bridge 49, 52
Samurai 77, 112
Samy Vellu 53
Sands, Bobby 101
Saratoga 105
Scargill, Arthur 101
Schutztaffel 44
Scud missiles 36
Seppuko 111
Severn Bridge (first) 47
Sewage effluent 29
Sharon, Ariel 74, 98, 119
Shatt al-Arab waterway 74, 75
Sherpas 33,34
    Nima Chhamzi Sherpa 34
    Sherpa language 34

Tensing Norgay 31
Shipton, Eric 33
Sinai Peninsula 98
Sinn Féin 65, 101
Sistine Chapel 12
Slave trade 38, 96, 105, 106
Snaefellsjökel 41
Somerville College, Oxford 99
South Georgia 101
Soviet Union 23, 62, 69-74, 90,
    96, 103-104, 111
Sparrowhawks 86
'Spirit of St. Louis' 23
Spitfires 37
Staircases 1, 5
Stanley, Sir William 34
Stannaries 41, 48
State of Israel 74, 98
Stephen, Feast of 55
Stimson, Henry L. 111
Sudetenland 38, 65, 66
Suez Canal 60, 73, 74, 98
Superbus, Lucius Tarquinus 46
Sur le Pont d'Avignon' 55
Suurhusen, Leaning Tower of 6
Tagore, Rabindranath 110
Tea Tax 106
Thatcher 99, 102, 103, 105
    Carol 103
    Denis 99, 102, 103
    Margaret 99, 100-103,
    Mark 102, 103
Taugwalders 31
Tavistock 48
Tensing Norgay 33
Thackeray, William M. 85

'The Last Judgement' 12
The London Hospital 5, 31, 49
Theodoric 79
Thermopylae 46
Tiberias 98
Tigris, River 74
Tojo, Hideki 111
Trebuchet 4
Trollope, Anthony 85
Tulagi 111
Uffizi Palace 16
Ukraine 104
United Nations Security Council
    101
Vasari, Giorgio 16
V-1 flying bomb 25, 36
V-2 rocket 37, 38
Verne, Jules 41, 59
Vicksburg 4
Virginia 105, 106
Visigoths 79
Wagner, Richard 87
Walking the plank 55
Wallis, Barnes 26
Wanaka, NZ, Leaning tower of 7
Wannsee Conference of 45
Washington, George 105, 106
Waterloo 83, 85, 101
Watt, James 107
Wedgewood, Josiah 107
Wellesley, Arthur 83, 84
Wellington, Duke of 83-86
Wenceslaus 54, 55, 71
Western Cwm 31
Wharfedale 30
Whitehouse, Mary 13

Wiesenthal, Simon 45
Wings, shape of 13, 21, 25, 26
World War II 6, 8, 27, 28, 35-37,
    42, 46, 54, 6-3, 112
Wilson, Maurice 33
Wind tunnel 21
World Trade Center 3
Whymper, Edward 31, 32
Wright brothers 21
Xerxes 47
Yeltsin, Boris 104, 105
Yom Kippur 15, 16
Zeppelins 19
Zermatt 30-32
Zhaozhou 46